Sustainable Strategic Business Infrastructure Development and Contemporary Digital Practices in Industry 5.0

Sustainable Strategic Business Infrastructure Development and Contemporary Digital Practices in Industry 5.0

Edited by

Indrajit Ghosal, PhD

Bikram Prasad, PhD

Nilanjan Ray, PhD

Lalit Kumar Joshi, PhD

AAP | APPLE ACADEMIC PRESS

First edition published 2025

Apple Academic Press Inc.
1265 Goldenrod Circle, NE,
Palm Bay, FL 32905 USA

760 Laurentian Drive, Unit 19,
Burlington, ON L7N 0A4, CANADA

CRC Press
2385 NW Executive Center Drive,
Suite 320, Boca Raton FL 33431

4 Park Square, Milton Park,
Abingdon, Oxon, OX14 4RN UK

Apple Academic Press exclusively co-publishes with CRC Press, an imprint of Taylor & Francis Group, LLC

Library and Archives Canada Cataloguing in Publication

CIP data on file with Canada Library and Archives

Library of Congress Cataloging-in-Publication Data

CIP data on file with US Library of Congress

ISBN: 978-1-77491-626-1 (hbk)
ISBN: 978-1-77491-627-8 (pbk)
ISBN: 978-1-00349-216-0 (ebk)

About the Editors

Indrajit Ghosal, PhD
Associate Professor, Brainware University, West Bengal, Kolkata (India)
Dept: School of Management & Commerce

Indrajit Ghosal, PhD, presently serves as an Associate Professor at Brainware University, Barasat, Kolkata, India, School of Management and Commerce. Dr. Ghosal was awarded a PhD in Online Marketing (Management) in 2020. He was awarded FRPM in 2017. He earned his masters in MCA and undergraduate with BSc.IT. He has more than 16+ years of experience in teaching. He has published 60 research papers in many reputed journals with indexing such as ABDC, ABS, Scopus, WOS, UGC, and many more, and published 6 patents. He has presented papers in 36 International, 7 national, and 4 case studies in many conferences. He is the author of 6 Books and serves as an editor in 11 Books that would be published (CRC- Taylor & Francis, Emerald, Wiley, Routledge, Bentham Science, Nova Science, etc.) and published 8 Book chapters. He assists 4 Ph.D. scholars, three MPhil scholars already awarded under his supervision and 1 scholar is pursuing. He serves as a reviewer in many reputed journal indexes in Scopus, WOS, ABDC, ABS, and SSCI. As a Keynote Speaker/ Resource person /Chairperson he has been invited to 30 Conferences/workshops/seminars. He is a Member of ACM (Meerut Chapter), SSRN, MTTF (Bhatinda, Punjab), Life Member and executive member of the Education Research and Development Association (ERDA), Member of IARA (Guwahati), IARA (Tiruchirappalli) & World Leadership Academy. His areas of research are Digital Marketing, Consumer Behaviour, Online Marketing, Entrepreneurship, E-commerce.

Bikram Prasad, PhD

Assistant Professor, Dr. B. C. Roy Academy of Professional Courses (Formerly Dr. B. C. Roy Engineering College), Durgapur

Bikram Prasad, PhD, is presently working as an Assistant Professor in the Department of Hospital Management at the School of Management, Swami Vivekananda University, India. He has five years of academic experience. He graduated as a pharmacist after completing a Bachelor of Pharmacy. He has also completed his postgraduation in Hospital Administration (Master of Hospital Administration). His core areas of interest are marketing management, data analytics, e-commerce, pharmacy management, medication management, hospital management, public health management, entrepreneurship, frugal innovation, total quality management, and consumer behavior. He has presented 14 papers at national and international conferences and published 11 research papers in reputed refereed, peer-reviewed journals from reputable publishers with indexing from Scopus, WOS, UGC, and many more. He is the author of one book and has published one book chapter in 2020. He has published one edited and one authored book in 2022.

Nilanjan Ray, PhD

Associate Professor of Marketing Management Department of Management Studies, JIS University, West Bengal, India

Nilanjan Ray, PhD, is from Kolkata, India, and presently associated as an Associate Professor at the Department of Management Studies, JIS University, India. Prior to joining JIS University, Dr. Ray was at the Institute of Leadership Entrepreneurship and Development as Associate Professor and Department Head and with additional responsibility as Director IQAC before that he was also associated at Adamas University as Associate Professor of Marketing Management and the Centre Coordinator for Research in Business Analytics in the Department of Management, School of Business and Economics, West Bengal, India. Dr. Ray obtained a certified Accredited Management

Teacher Award from the All India Management Association, New Delhi, India. He also earned various degrees: PhD (Mktg); M.Com (Mktg); MBA (Mktg), STC FMRM (IIT-Kgp). He has more than 12 years of teaching and six years of research experience, awarded 3 Doctoral Scholars, and guided around 56 postgraduate students' project also. Dr. Ray has contributed over 90 research papers in reputed National and International Referred, Peer Reviewed Journals, Proceedings and 13 Edited Research Hand Books from Springer, IGI-Global USA and Apple Academic Publisher CRC Press (A Taylor & Francis Group), USA. He has obtained 1 Patent from Germany and 2 Copyright from India. He has also associated himself as a reviewer of Tourism Management Elsevier, Journal of Service Marketing Emerald Group Publishing Limited, Journal of Business and Economics, Research Journal of Business and Management Accounting, and as an Editorial Board Member of several referred Journals. Dr. Ray has organized several FDPs, National and International Conference, and Management Doctoral Colloquium. Dr. Ray is a life member of Fellow Member of Institute of Research Engineers and Doctors Universal Association of Arts and Management Professionals (UAAMP) New York, USA Calcutta Management Association (CMA).

Lalit Kumar Joshi, PhD

Assistant Professor, Department of Commerce at Kazi Nazrul University, West Bengal, India

Lalit Kumar Joshi, PhD, is an Assistant Professor in the Department of Commerce at Kazi Nazrul University, West Bengal, India. He has over 14 years of teaching experience, of which 10 years are at the postgraduate level. Before joining Kazi Nazrul University, he was a lecturer at The Bhawanipur Education Society College, Kolkata, W.B., India. Dr. Joshi earned his PhD degree in 2017 from Vidyasagar University in the Indian Steel Industry, where he was the recipient of gold medals for his outstanding performance in the MCom Examination. His specialization is in the area of banking and finance. He has published several research articles in reputed national and international journals and has presented several research papers at national and international conferences. He has edited three books and has served as an editor of *The BESC Journal of Commerce and Management.* He has also authored books titled *Receivables Management: An Empirical Study with Reference to Selected Companies in the Indian FMCG Industry* and *Marketing Management and Human Resource Management (Text and Cases).*

Contents

Contributors .... *xi*

Foreword .... *xv*

Preface .... *xvii*

1. **Analyzing the Potential of Bitcoin to Become a Global Currency 1**
Satyam Kumar and Kalpana Singh

2. **Industry 5.0: A Solution for Leading SMEs 21**
Monika Arora, Prema Mahale, and Kuldeep Bhalerao

3. **Strategies for Channelizing Women Artisans as Entrepreneurs to Leverage the Indian Economy: An Empirical Study of MSME (Handicraft Sector) Concerning ODOP in Uttar Pradesh 33**
Uma Shankar Yadav, Ravindra Tripathi, and Mano Ashish Tripathi

4. **Managing (Mis-/) Disinformation Through Big Data Analysis: Resorting to Sustainable Strategies to Counter Infodemic 57**
Manpreet Arora, Roshan Lal Sharma, Vaishali Dhiman, and Sunaina Rathore

5. **Impact of COVID-19 on Vegetable Sales and Consumption and Its Effects on the Health Sector in Kathmandu for Sustainable Management 73**
Indrajit Ghosal, Meen Bahadur Karki, and Mahima Karki

6. **Review of Industry 5.0: Evolution and Future Advancements 87**
Pankaj Anand, Jitendra Kumar, and Subhadeep Mukherjee

7. **Exploring Cloud Computing Platforms and the Metrics Used to Compare Cloud Service Providers 107**
Parul Dubey and Arvind Kumar Tiwari

8. **India: The Digital Economy and Its Financial Inclusion with Special Reference to Industry 5.0 .... 125**
Piyali Roy Chowdhury and Indrajit Ghosal

9. **An Empirical Study on Accounting Conservatism, Integrated Reporting in India, and Corporate Innovation: Evidence from Industry 5.0 .... 139**
Farida Rasiwala and Indrajit Ghosal

10. **Assessment of Investment Priorities by Rural Women 161**
Koppala Venugopal

11. **Evaluating the Impact of Physical and Digital Infrastructure for the Development of Tourism Business in North-East India, with Special Reference to Assam 175**
Soumendra Nath Biswas and Puspanjali Mohapatra

12. **Firm Types and Design Processes: A Typological Analysis in the Indian Context 185**
Santanu Roy and Parthasarathi Banerjee

13. **Wearable Technology: Developments, Business Models, and Current Challenges 201**
Priyanka Singh and Harit Kumar

14. **Motivations, Key Success Factors, Functionality, and the Role of Social Entrepreneurship in Socio-Economic Sustainability: A Systematic Review of Existing Literature 217**
Arun Mittal, Priyank Kumar Singh, and P. Horsley Solomon

15. **Healthcare Industry 4.0 to 5.0: A Psychological Study 237**
Binda Kumari

16. **Analysis of Sustainable Development in Spatial Politics in the Construction of the Raknamo East Nusa Tenggara in Indonesia 255**
Melkisedek N. B. C. Neolaka, Diana San Tabun, and Uma Shankar Yadav

17. **The Impediments to Student Life Imposed by COVID-19 in India and Bangladesh: An Empirical Analysis 271**
Dhanonjoy Kumar, Nirmal Chandra Roy, and Nilanjan Ray

18. **Effectiveness of Social Media Marketing as a Sustainable Futuristic Tool for Consumer Decision-Making Process 289**
Ahasanul Haque and Khushbu Agarwal

19. **Framework for Adopting a Data Analytics Strategy in the Fourth Industrial Revolution as a Stepping Stone to the Fifth Industrial Revolution 301**
Garikayi Emmanuel Gondo, Lorna B. Holtman, and Ravinder Rena

Index 317

Contributors

Khushbu Agarwal
Pacific Institute of Management, Pacific University, Udaipur, Rajasthan, India

Pankaj Anand
Research Scholar, Department of Management Studies, Sikkim Manipal University, Sikkim, India

Manpreet Arora
Assistant Professor, HPKVBS, School of Commerce and Management Studies, Central University of Himachal Pradesh, Dharamshala, Himachal Pradesh, India

Monika Arora
Professor, Amity Business School, Amity University, Manesar, Haryana, India

Parthasarathi Banerjee
International Management Institute, New Delhi, India

Kuldeep Bhalerao
Assistant Professor, Bharati Vidyapeeth's Institute of Management Studies and Research, Navi Mumbai, Maharashtra, India

Soumendra Nath Biswas
Associate Professor and Head, Department of Hospitality and Tourism Management, Assam University (A Central University), Silchar, Assam, India

Piyali Roy Chowdhury
Assistant Professor, Institute of Management Study, West Bengal, India

Vaishali Dhiman
Research Scholar, HPKVBS, SCMS, Central University of Himachal Pradesh, Dharamshala, Himachal Pradesh, India

Parul Dubey
Assistant Professor, Department of Information Technology, Shri Shankaracharya Institute of Professional Management and Technology, Raipur, Chhattisgarh, India

Indrajit Ghosal
Brainware University, West Bengal, Kolkata, India

Garikayi Emmanuel Gondo
UNISA School of Business Leadership (SBL), University of South Africa (UNISA), Midrand, South Africa

Ahasanul Haque
Professor of Marketing, Department of Business Administration, International Islamic University Malaysia, Kuala Lumpur, Malaysia

Lorna B. Holtman
Honorary Research Fellow, International Education and Partnerships, Durban University of Technology, ML Sultan Campus, Durban, Republic of South Africa

Mahima Karki
School of Management, Tribhuvan University (SOM TU), Kathmandu, Nepal

Meen Bahadur Karki
Research Scholar, School of Management and Commerce, Poornima University, Jaipur, Rajasthan, India

Dhanonjoy Kumar
Professor, Department of Management, Islamic University Kushtia, Bangladesh

Harit Kumar
Associate Professor, Pranveer Singh Institute of Technology, Kanpur–Agra–Delhi National Highway-2, Bhauti, Kanpur, Uttar Pradesh, India

Jitendra Kumar
Associate Professor, Department of Management Studies, Sikkim Manipal University, Sikkim, India

Satyam Kumar
BBA VI SBS

Binda Kumari
Head of Department, Department of Psychology, K.O.C., Ratu Ranchi (Ranchi University), Ranchi, Jharkhand, India

Prema Mahale
Assistant Professor, VESIMSR, Chembur, Mumbai, Maharashtra, India

Arun Mittal
Assistant Professor, Birla Institute of Technology, Mesra, Ranchi, Off-Campus Noida, Uttar Pradesh, India

Puspanjali Mohapatra
Utkal University of Culture, Bhubaneswar, Odisha, India

Subhadeep Mukherjee
Ramaiah University of Applied Sciences, Bengaluru

Melkisedek N. B. C. Neolaka
Faculty of Social Science and Political Science, University of Nusa Cendana, NTT, Indonesia

Farida Rasiwala
Fortune Institute of International Business, New Delhi, India

Sunaina Rathore
Research Scholar, HPKVBS, SCMS, Central University of Himachal Pradesh, Dharamshala, Himachal Pradesh, India

Nilanjan Ray
JIS University, West Bengal, India

Ravinder Rena
DUT Business School, Faculty of Management Sciences, Durban University of Technology, ML Sultan Campus, Durban, Republic of South Africa

Nirmal Chandra Roy
Assistant Professor, Department of Business Administration (HR), The University of Burdwan, West Bengal, India

Santanu Roy
ICFAI Business School, Dehradun

Roshan Lal Sharma
Professor, Department of English, Central University of Himachal Pradesh, Dharamshala, Himachal Pradesh, India

Kalpana Singh
Assistant Professor, Sharda School of Business Studies, Sharda University, Greater Noida, Uttar Pradesh, India

Priyank Kumar Singh
Doon University, Dehradun, Uttarakhand, India

Priyanka Singh
Assistant Professor, Pranveer Singh Institute of Technology, Kanpur–Agra–Delhi National Highway-2, Bhauti, Kanpur, Uttar Pradesh, India

P. Horsley Solomon
Head, Department of Electronics Science, SRM Arts and Science College, Kattankulathur, Chengalpattu, Tamil Nadu, India

Diana San Tabun
Faculty of Social Science and Political Science, University of Nusa Cendana, NTT, Indonesia

Arvind Kumar Tiwari
Professor and Dean Academic, Dr. C. V. Raman University, Chhattisgarh, India

Mano Ashish Tripathi
Department of Humanities and Social Sciences, Motilal Nehru National Institute of Technology, Prayagraj, Uttar Pradesh, India

Ravindra Tripathi
Department of Humanities and Social Sciences, Motilal Nehru National Institute of Technology, Prayagraj, Uttar Pradesh, India

Koppala Venugopal
Professor, Department of MBA, Aditya Institute of Technology and Management, Tekkali, Srikakulam, Andhra Pradesh, India

Uma Shankar Yadav
Department of Humanities and Social Sciences, Motilal Nehru National Institute of Technology, Prayagraj, Uttar Pradesh, India

Foreword

It gives me immense pleasure and pride to introduce and write the foreword for the book *Sustainable Strategic Business Infrastructure Development and Contemporary Digital Practices in Industry 5.0.*

For the first time in most of the world's history, during the Covid-19 pandemic, the majority of the populace were spending time indoors against their free will and as a result of government regulations. Everyone, regardless of age, including children, teenagers, and young adults from all age groups, were ordered to stay indoors. Smartphones and computers were people's best friends, sources of entertainment, and means of communication and information, especially in the early years of social distancing, even though newspapers weren't printed or disseminated. Of course, this is the tale of the middle- and upper-middle classes.

During this, dependency on phones grew multi-fold. We all started participating in and creating indoor programs for recitation, musical extravaganzas, cinematic workouts, etc. A new experiment in indoor living began, and to be completely honest, the majority of the population was able to make up for the lack of the outside world by utilizing the Internet's virtual environment. The majority of rural India was excluded from this forced indoor anguish.

After the pandemic lockdown, another important development occurred. The employers and the leaders of our central and state governments in India abandoned thousands of our own countrymen who had traveled far and wide during normal times to make a living in various regions of our enormous country. To get home, they traveled hundreds of kilometers. There was a veritable flood of people on the railroad tracks and highways. Mobile phones were their only means of communication with their loved ones. Some of them were really clever, using Google Maps to navigate the maze of railroad tracks.

The aforementioned data indicates that digitalization is now inextricably linked to our economy and society. Therefore, if we want to comprehend how it affects our people, studying our culture and economy is not only required but essential. Our entire population, including youngsters who are three years old and older, adolescents, young adults, middle-aged individuals, and senior citizens, both male and female, are all connected via WhatsApp, Facebook, Messenger, Twitter, and various other connecting gadgets. The fact that these devices are interactive is their most important feature. Unlike radio and television, these apps provide a venue for anybody and everyone to express themselves and hear other people's opinions.

Before the invention of the telephone, radio, and television, we would remark that the entire world had transformed into a hamlet, or Basudhaiba Kutum-Bakam. Today, anything can be done on a laptop, a computer, a smartphone, or a tablet. The entire world is literally at our fingertips. Since the material they provide is no longer updated, printed encyclopedias have been eliminated. Today, no guru is sacred because information can be found instantaneously on a smartphone. On request, textbooks and reference materials are available. On the internet, you can get a variety of translations of the world's literature in a variety of languages. On the mobile, you may get information on any issue, including COVID-19 and the most recent news regarding the last world war. With just a click, you can learn about the scenic beauty and political or socioeconomic problems of Fiji, Chile, Vietnam, or Madagascar. The sixth wonder of the world is this complex, multilayered device that can fit in your pocket or be held in your palm; however, we no longer view it as a wonder due to its easy accessibility. Such an effort to comprehend this phenomenon is made in this book.

A total of 13 chapters covering significant subjects have shed light on the topic at hand: Chapters on management sciences, essays on gender studies and human behavior, big data analysis, industry 5.0, and the digital economy have been included in this edited volume. To justify the utilitarian objective, the chapters regarding cloud computing and healthcare 5.0, corporate innovation, digital infrastructure of the business, and design process of forms have been included in this edited volume.

We are pleased that this book has been published. We anticipate the publication of more volumes covering a wide range of different facets of the digital world. The digital divide is a significant issue. How this could be minimized should be carefully considered. The world of the future will be both physical and digital.

—Sudhir Rana, PhD
Associate Professor of Marketing and Strategy,
College of Healthcare Management and Economics,
Gulf Medical University, UAE

Preface

Businesses will experience change as the world unites to accomplish the Sustainable Development Goals of the United Nations. To align with sustainability goals that support industrial innovation, uphold wellness, safeguard the environment, and guarantee the creation of commercial value, firms must reevaluate their strategies and business practices and evaluate their impact. Strong dedication and the tools to put this ambitious agenda into action are needed. This volume on technical innovations for sustainability and corporate growth is an essential source of research on how to speed up business growth through technological innovations and knowledge generation.

Numerous current trends demonstrate that the modern world economy is on the verge of a new industrial revolution. First, the early 21st century's protracted global economic crisis and the inability to resolve it using current economic system options demonstrate that the prior technological model's potential has been exhausted. In the area of industrial production, the crisis first appeared as an overabundance of industrial goods and their inability to be sold in domestic economic systems or on international markets, which resulted in widespread corporate failures throughout the world and an uptick in protectionist measures from governments of various nations. Second, ending the global crisis necessitates beginning a new wave of innovations, according to contemporary economic theory (specifically, the theory of economic cycles, the theory of crises, the theory of innovations, etc.). This trend is encouraged by the rapid development of the knowledge economy in many nations, which has increased the potential for future innovation in the global economic system. A widely acknowledged global priority of socioeconomic growth is innovation. Third, throughout the past few decades, researchers from various nations have carried out research that has led to the development of new technologies, the majority of which are leading production technologies (technological innovations); that is, they are targeted at the real sector of the economy.

An international initiative to gather multi- and interdisciplinary research that best supports innovation has culminated in this book. This prelude provides a succinct summary of recent organizational and innovative advances.

A comprehensive view and analysis of innovation at the level of business operations, including advances in HR, is provided by sustainability, technology, and innovation 4.0. Its chapters as a whole provide the concept that innovation 4.0 denotes a technological revolution framed within the path of sustainable development. This edited book accomplishes methodical, empirical, and utilitarian goals by providing a thorough and comprehensive assessment of Innovation 4.0. The methodological goal is to provide tools that make it possible to identify, analyze, and evaluate the relationship between innovation 4.0 and inspiration that will lead society toward a new economic and social order.

Its empirical goal is to make it possible to analyze and assess how public organizations, industrial innovations, and HR innovations contribute to the process of creating sustainable development of the global environment. The book's utilitarian objective is to suggest Innovation 4.0 as a tool for international organizations to promote an inventive economy.

This book is given in this context to offer thorough assistance for Industry 4.0 applications. As a result, this book presents a conceptual framework for Industry 4.0 with respect to its design principles in addition to introducing the implementation aspects of Industry 4.0. A maturity and readiness model is also suggested so that businesses choosing to go down the route of digital transformation can assess themselves and solve the issue of determining where to start. A technology roadmap is also provided to show managers how to set Industry 4.0 strategies, choose key technologies, decide on projects, build an optimized project portfolio while taking risk into account, and schedule projects over the course of a planning horizon. Also highlighted are the effects of digital transformation on engineering education and talent management. The book then moves on to important technological developments that serve as the foundation for Industry 4.0, exploring their potential technical and economic benefits through examples drawn from actual applications.

CHAPTER 1

Analyzing the Potential of Bitcoin to Become a Global Currency

SATYAM KUMAR[1] and KALPANA SINGH[2]

[1]*SSBS, Sharda University, Greater Noida, Uttar Pradesh, India*

[2]*Assistant Professor, SSBS, Sharda University, Greater Noida, Uttar Pradesh, India*

ABSTRACT

Globalization has brought along countless innovations that have become crucial to modern society. Amongst these is the introduction of crypto currencies. Despite the reservations voiced by many regarding their viability, crypto currencies continue to be strong candidates for what could become a global currency. This chapter examines the probability of Bitcoin, the first decentralized crypto currency, becoming the shared currency of the world. The research aims to identify and analyze the qualities and conditions necessary for the emergence of a global currency, as well as Bitcoin's potential as a candidate. The results of this study will shed light on the prospect of Bitcoin becoming the global currency of the future.

1.1 INTRODUCTION

The modern era of globalization has brought about a significant level of integration, standardization, and uniformity to the world. Over the past 200 years, humanity has witnessed various advancements, including the

Sustainable Strategic Business Infrastructure Development and Contemporary Digital Practices in Industry 5.0. Indrajit Ghosal, Bikram Prasad, Nilanjan Ray, & Lalit Kumar Joshi (Eds.)

establishment of crucial global organizations such as the United Nations and the World Trade Organization.

Representatives from different countries now gather more frequently to coordinate policies and remove barriers. As the world becomes more interconnected and global trade flourishes, the need for a shared global currency should emerge as the next pivotal step for the world economy. This would eliminate the risks associated with fluctuating foreign exchange rates, reduce transaction costs, and facilitate an even greater volume of international trade in goods and services.

A global currency can be defined as a medium of exchange that is shared and used globally. As the currency that will be shared by the worldwide population, it should possess at least two qualities:

1. **High Liquidity:** The ease to exchange such currency for goods and services despite geographical location.
2. **Low Risk:** Low sensitivity towards volatile market movements.

Being the regionally shared currency of Europe, the European Euro serves as a precedent and important case study for the establishment of a global currency.

The success of the European Euro should lay the foundation for the impending global currency. In the late 1960s, it was decided that the European Union (EU) would provide "stability and an environment for growth and development" ("The History of the Euro," 2017). Over a decade (1990-1999), the EU completed three economic preparatory steps to introduce the newly created shared currency, the Euro: establishing an internal market with free capital movements within the region, founding a regional central bank to regulate the new currency, and fixing the exchange rate and launching the currency ("The History of the Euro," 2017). Born from significant effort and collaboration, the Euro has offered both advantages and disadvantages to the region. As the shared currency of 19 out of 28 EU members, the Euro has become a steadfast commitment that connects the economies of 19 different countries. Even though the EU has made great effort in the preparatory phase for the Euro to create an initial internal market, those member countries that joined in later sometimes brought along unbridgeable gaps that skewed the balance of the regional economy.

Nonetheless, the shared Euro also brought countless advantages that proved to have outweighed those shortcomings previously discussed. These benefits include the elimination of fluctuating exchange rates and exchange costs, as well as the stimulation of free trade on a regional scale.

From this brief history of the Euro, we can establish three conditions for successfully implementing a globally shared currency:

- First, the countries that join the shared currency union must be at the same (or at least similar) levels of economic development.
- Secondly, the chosen currency must be politically neutral. In other words, the shared currency cannot be the currency of any member countries, as this would greatly skew the economic power balance in favor of that particular country.
- Third, there should be a collection of entities (governments in this case) to enforce and regulate the functioning of the new currency.

Currently, there are three candidates for the role of the global currency: the traditional gold standard, one of the prevailing major currencies, or a representative from the emerging cryptocurrencies. Gold has been the core of all exchange mediums since the end of the bartering era. The centuries of its existence have built an unwavering trust in people regarding its value; and in turn, this unwavering trust lies at the core of gold's persisting value. Consequently, gold possesses the quality of a safe-haven, a traditional first-choice reservoir of wealth whenever there is instability or fluctuation in the financial market. Considering the two required qualities mentioned above for the global currency, gold seems to fit the bill perfectly. It has great liquidity as its value is acknowledged all over the world (thus can be converted into any currency), and it has the best safety insurance due to people's unchanging trust in its value. Nonetheless, gold fails to satisfy any of the three implementation conditions for the global currency. To be more specific, gold mines are distributed unequally all over the world and major countries tend to have much greater gold reserves compared to others (the perk of being in a higher position in international trade for centuries). These facts make it impossible for gold, although ubiquitous all over the world, to be adopted as the global currency. Gold is not a politically neutral currency, which would greatly favor the current major countries in the world. Additionally, there is no existing entity that can regulate the use and distribution of gold. Furthermore, the difficulties in its daily usage, management, and distribution are decisive reasons why gold becomes obsolete and less favorable as the world's currency. History has proven that convenient paper money and online credit will eventually replace the need for gold in daily trade.

Another possible candidate for the role of global currency is one of the current major currencies. The United States Dollar, the European Euro, the Japanese Yen, and the British Pound are currently the four strongest currencies. Among these, the U.S. Dollar stands out as the previous standard reserve asset in the Bretton Woods system and the current important standard peg of various currencies. It is widely used as an intermediary currency in foreign exchange, making it the most likely candidate. However, choosing the U.S. Dollar as the global currency would concentrate even more economic and political power at the expense of the United States, which is unlikely to gain the consensus of other countries.

As an alternative, this study examines the possibility that cryptocurrencies, represented by Bitcoin, could become the global currency. Bitcoin is a strong contender for this position because it possesses the qualities required for a future global currency. Firstly, Bitcoin exhibits a high level of liquidity as it is widely used as an intermediary currency between major currencies such as the U.S. dollar, Euro, Japanese yen, Chinese RMB, and others. This is due to its ease of conversion and reduced paperwork and transactional fees. Secondly, extensive research has confirmed that Bitcoin shares similar characteristics with safe-haven assets like gold and the U.S. dollar, making it a suitable option for wealth storage. Additionally, Bitcoin satisfies the three conditions necessary for successful implementation as a global currency. Firstly, it primarily targets the online payment market segment, with major Bitcoin traders being developed countries with similar economic conditions such as the United States, Japan, and the European Union. Secondly, being a decentralized currency, Bitcoin is politically neutral compared to other candidates. It has no direct affiliations with any central authority, which promotes a free market and equal impacts on individual countries when adopted as the global currency. Furthermore, Bitcoin's relatively recent emergence, only eight years ago, ensures that no country has accumulated enough Bitcoin reserves to become the most influential figure in the Bitcoin market. Lastly, despite being decentralized, Bitcoin does have a collection of entities, known as miners, that regulate its functioning.

1.2 LITERATURE REVIEW

1.2.1 BITCOIN: A BRIEF HISTORY

On October 31st, 2008, Satoshi Nakamoto set the foundation for Bitcoin by publishing a white paper through metzdowd.com, a cryptography mailing list, to introduce the technical framework of Bitcoin and the solution to prevent cryptocurrency's problem of double-spending ("Bitcoin History"). On January 3rd, 2009, the first Bitcoin blockchain, known as the genesis block, was mined ("Bitcoin History"). From June 2011 until now, Bitcoin prices have experienced several crashes, which have contributed to Bitcoin's notoriety in public opinion ("Bitcoin History"). However, in the article "Anniversary of the Great Bubble of 2011," Buterin, a co-founder of Bitcoin Magazine, claimed that during the post-bubble period, the Bitcoin community acquired most of its current members, and the fast recovery demonstrated that the members of the Bitcoin community are willing to endure hard times for the sake of a brighter long-term future (2012).

This endurance exhibited by the Bitcoin user community is the deciding factor for Bitcoin's stability and trustworthiness as a currency. The value of every currency is based on the conventions established by people: gold has such strong value due to the longevity of the gold standard convention, and the dollar, despite being a fiat currency for quite some time, has such strong value because the public believes that the U.S. government will vouch for its value through the ups and downs of the economy. Bitcoin is no different: with unwavering trust from the Bitcoin enthusiast community and their experience with various fluctuations in Bitcoin's history, the fact that they voluntarily remain as Bitcoin users proclaims their steadfast trust in the future major role that Bitcoin will play in the global market. This convention, based solely on speculation, may rival or even surpass the conventions enforced by central authorities.

1.2.2 GROUNDWORK SET BY SATOSHI NAKAMOTO'S "A PEER-TO-PEER ELECTRONIC CASH SYSTEM" (2008)

In his paper published in 2008, Nakamoto proposes the Bitcoin system that disruptively innovates the way electronic transactions are initiated and processed. He argues that the traditional method of processing transactions through a trusted third-party, usually financial institutions, always suffers from the "inherent weaknesses of the trust-based model" as banks cannot offer a completely non-reversible transaction in case of disputes. As reversibility remains, the need for trust increases. He concludes that by providing a cryptographic proof instead of relying on trust, Bitcoin can become the first transactional system that allows people to make direct electronic exchanges without the need for a trusted third party, an independent system for online peer-to-peer transactions. The major setback to the realistic implementation of previous cryptocurrencies is the unresolved problem of possible double spending. Nakamoto directly addresses this problem in his paper and Bitcoin's framework: by implementing a timestamp in each blockchain and keeping a public history record of transactions, the honesty of each block will be confirmed or rejected by the majority vote of CPU power, the network of computer power provided by Bitcoin miners and users. He also further states that any future necessary rules or incentives can be decided using the same "consensus system."

1.2.3 CHARACTERISTICS OF BITCOIN

Theoretically, there is no limit to Bitcoin's liquidity. Being a peer-to-peer online payment system, Bitcoin can be exchanged for goods, services, or currencies as long as both traders agree to such transactions. In reality, this ability of Bitcoin is greatly limited due to its low rate of exposure (or popularity). Cryptocurrency is a novel concept; thus, the risk associated with it can be considered unreasonable to those who advocate the use of traditional currencies. However, Bitcoin's superiority in transactional speed and reduced fees will surely contribute to its growing popularity, allowing people to understand its mechanism more thoroughly and gradually accept the innovative convenience it brings.

Currently, Bitcoin has been playing a significant role as an intermediary for various major currencies: Japanese Yen, United States Dollar, Euro, Korean Won, and British Pound. However, as the Bitcoin community continues to grow, the potential liquidity of Bitcoin should be recognized and valued in the future.

In a research by Bouri, Jalkh, Molnar, and Roubaud (2017), the potential properties of Bitcoin as a diversifier, hedge, and safe-haven were examined. The study used the pair wise dynamic conditional correlations (ADCC model) to analyze the relationship between Bitcoin and three commodity indices: general commodity index, energy commodities index, and non-energy commodities index. The analysis was conducted for two sub-periods, before and after the December 2013 bitcoin crash.

The results of the analysis indicate that Bitcoin has a "time-varying diversification ability" and can serve as a "strong hedge and safe-haven for energy commodities, but not for non-commodities." However, it is important to note that this tendency is weaker compared to the safe-haven property of gold, which is globally recognized.

Furthermore, it should be considered that the results may have been influenced by the limited participation of the global financial market in Bitcoin trading. Currently, only a small portion of the market engages in Bitcoin transactions, resulting in low sensitivity to market volatility. As more traders join the Bitcoin transaction system, it is expected that Bitcoin will reflect market volatility more closely.

Nonetheless, these facts are also insufficient to claim that Bitcoin is an unmanageable risky currency. In the eight-year period of its existence, Bitcoin has been through countless ups and downs, leading the majority of the public to often doubt the safety that Bitcoin guarantees. In response to this issue, there are two important facts to be noted. First, Bitcoin is still in its infancy stage when compared with other more "mature" currencies. Therefore, it would only be fair to compare Bitcoin with other currencies when it is given more time to grow and stabilize. Despite its volatility, Bitcoin has consistently rebounded and gained more popularity, participation, and faith from the people. In fact, the overall trend for Bitcoin shows exponential growth, to the point where some critics fearfully predict inflation and speculation bubbles. In the following subsections, the reasons why such fears are baseless will be discussed in detail.

1.2.3.1 IMPLEMENTING CONDITIONS 3.2.1. THE SHARED MARKET

One condition for the successful implementation of a shared currency is that the member countries must have the same level of economic development. This condition would be automatically fulfilled if Bitcoin were chosen as the global currency. Bitcoin is a currency system designed for online payments only. The history of world economic development can be divided into different periods or stages based on the prevailing method of payment: starting from barter, gold, silver, bronze coins, banknotes, paper money, debit/credit, and now Bitcoin. Currently, all the countries that have embraced Bitcoin as a major form of trade are developed countries where online payment is the primary method of purchase. As a new union with a shared currency (Bitcoin) is established, these countries (at the same stage of economic development) would become the initial members, forming the initial internal market that would support the shared currency, until other countries can also progress enough to adopt online payment as their primary method of payment.

1.2.3.2 POLITICAL NEUTRALITY

Bitcoin is a decentralized currency: it does not involve any governments, central or local banks in its workings. Consequently, when chosen as the global currency, Bitcoin will not skew the economic balance towards any political powerhouse.

1.2.3.3 REGULATIVE BODY

Despite being decentralized, Bitcoin still has a collection of entities to regulate its mechanism: the group of miners and developers. Utilizing the majority consensus method provided by Satoshi Nakamoto when designing, Bitcoin gives the right and power to regulate it in the hands of the people, the community. Bitcoin is rather special in this case: it requires trust from no one and everyone at the same time. There is no need for a centralized trusted authority in the system because everyone in the community has access to the public records, and thus can judge the integrity of all that has been happening. The system took the trust away from one entity and put it in the hands of the community, trusting that with everyone watching and judging, everyone should be honest.

1.2.3.4 BITCOIN USER'S PROFILE

The number of current Bitcoin users is estimated to be around 15 million users. From research by Ermakova et al. the profile of a typical Bitcoin user has been created. According to the result of the survey, a Bitcoin user would typically be an adult of the working age from a developed country and possessing some knowledge of the financial market and technology advancements. Considering demographic data of the major countries that trade in Bitcoin (US, Japan, EU), the potential number of Bitcoin users should be at least 10-fold of that amount. One explanation to such disparity between real and potential number of users is Bitcoin's low rate of exposure. Bitcoin news appears more frequently on major and official newspapers and magazines only recently. In addition, one who knows of Bitcoin's existence may not necessarily know of Bitcoin's mechanism. After years of volatility, Bitcoin has gained much notoriety of instability. Newcomers of the Bitcoin community may be prejudiced before thoroughly getting to know Bitcoin.

1.2.3.5 THE FUTURE OF BITCOIN

1. **Hypothesis: Bitcoin as the Future Global Currency:** There are several pros and cons regarding Bitcoin's hypothesized future as a global currency.

   On the one hand, Bitcoin will bring various advantages to the global market. The first advantage lies in its distribution. Bitcoin is distributed through its own mining system: miners are rewarded with Bitcoin as they commit their CPU power to mine new Bitcoin. The cost of mining (or mining difficulty) will gradually increase over time, slowing down the mining rate until it eventually stops completely when all 21 million Bitcoins, the fixed amount of Bitcoins that will ever be mined and distributed, are available in the market. The purpose of this design is to simulate the rarity, the core value, of gold and to ensure that no inflation resulting from an over-injection of the currency into the money circulation can take place in the long run.

   The second advantage lies in its ease of use. Bitcoin's decentralized nature eliminates unnecessary paperwork, fees, and time consumption that result from the involvement of financial institutions. Therefore, Bitcoin users can easily complete a transaction in two steps: exchanging their currency for Bitcoin and setting up the transfer destination (a wallet address).

   On the other hand, Bitcoin also brings many concerns to potential users regarding its safety, regulation, and the possibility of deflation. In Ermakova et al.'s paper, safety has been identified as the leading challenge to the adoption of Bitcoin. Due to Bitcoin's notoriety in public opinion, many have failed to recognize its improvement from earlier years. Evidently, the number of theft attacks and price crashes has significantly decreased over time.

   Another concern that many have is the inflation resulting from Bitcoin's speculative bubble. The editors from The Economist have discussed this problem at great length in the article "What if the bitcoin bubble bursts?" and reached the conclusion that Bitcoin generated a "healthy bubble," as the system is still fairly "self-contained" and unlikely to spread contagion to the rest of the economy (2017). Additionally, as discussed above, the Bitcoin community is comprised of staunch believers; this characteristic adds to the non-contagious property of Bitcoin. As they are a community built on trust, the trust in the community and the risk endurance that they have accumulated up until now prevents the possibility of uniform withdrawals from the system.

The third often voiced concern is the fear of deflation. As the amount of Bitcoin is fixed at 21 million Bitcoin, when the number of Bitcoin reaches the ceiling limit, deflation is bound to happen: the price of Bitcoin will continuously rise in relation to commodities, increasing the incentive for savings and decreasing motivation for investment in the market. However, this criticism fails to take into consideration the divisibility of Bitcoin. As Cawrey mentions in his article on CoinDesk:

> "Bitcoin is currently divisible down to 0.00000001 of one BTC" (2013). This ability will become useful should the need arise to combat future deflation.

1.2.3.6 BITCOIN: A CLOSER FUTURE

Despite all the redeeming features that Bitcoin possesses to become the shared currency of the world, the possibility remains a prediction for years to come so that Bitcoin can further mature or better means of forecasting are devised. There is one thing, however, that can be predicted with much better precision – Bitcoin's short-term future. In the short term, along with the recent Bitcoin craze, trend, and newspaper exposure, Bitcoin should gain more and more new users and even higher market capitalization.

1.2.3.7 BITCOIN FUTURE PREDICTION – NEURAL NETWORK MODEL

1. **Purpose:** In order to learn and forecast the trend of the change in the number of Bitcoin transactions (or Bitcoin's popularity) in the coming year (2018), a machine learning neural network model has been chosen to make a general forecast using the data collected during the eight years that Bitcoin has been in circulation.
2. **Data Source:** Data in this research has been obtained from blockchain.info, a bitcoin cryptocurrency wallet and block explorer service. The studied dataset is the number of confirmed Bitcoin transactions for the period from 03 January 2009 to 15 November

2017 (from the issuance of the first Bitcoins until now) ("Number of Confirmed Bitcoin Transactions Daily"). The total number of observations is 1620. The data is collected and recorded every two days.

3. **Data Statistics:** Considering the significant skewness of the data (1.076) for the entire period from 2009 to 2017, the dataset will be divided into two subsets for two separate forecasting trials: one will use the full dataset, while the other will use a sample of data from January 2, 2015 to November 15, 2017 (525 observations) (Table 1.1).

TABLE 1.1 Dataset from 2009–2017 for Forecasting; blockchain.info (2009–2017–A Bitcoin Cryptocurrency Wallet)

Statistics	Full Dataset (2009–2017)	Sample (2015–2017)
Mean	83928.75494	206155.9295
Standard error	2408.650596	3150.962024
Median	51240.5	216812
Mode	125	203165
Standard deviation	96946.31638	72197.60992
Sample variance	9398588260	5212494879
Kurtosis	–0.129945494	–1.054788011
Skewness	1.07589616	–0.086754969
Range	367710	295476
Minimum	0	72234
Maximum	367710	367710
Sum	135964583	108231863
Number of observations	1,620	525

1.2.3.8 *THE MODEL*

The model is a machine learning neural network that has been customized using Python, various machine learning libraries (SkLearn, Pandas, NumPy, Matplotlib, and TensorFlow), and the source code from Aymeric Damien's Neural Network Example. The input dataset will be read from a file, normalized, and scaled (values 0–1) before any predictions are made. The program's parameters are set as shown in Table 1.2.

TABLE 1.2 Parameters of the Machine Learning Program

Parameters	Value
Number of epochs	4,000
Learning rate	0.01
Batch size	100
Hidden layer 1	30
Number of input (number of data for one year)	183
Number of output (number of data for one year)	183

The model will be used on the datasets twice. In the first trial, the full dataset will be used as the training set to predict the number of Bitcoin transactions in 2018. This trial is expected to yield poor forecasting performance due to the significant skewness in the dataset. In the second trial, the sample dataset will be used as the training set to predict the number of Bitcoin transactions in 2018. This trial is expected to produce better and more realistic results compared to the first trial, as the outlier data from the initial years has been removed from the training set.

The model will report the training loss and testing loss (mean square error of the data), a color-coded chart (blue line for the training data and orange line for predicted data), and an array of predicted data at the end of the Python program.

1.3 RESULT

The first trial produces a prediction with a train loss of 0.023035. Considering the entire dataset, the model predicts that the number of Bitcoin transactions in the year 2018 (183 observations) will increase beyond the current maximum. However, since the dataset is already normalized and scaled, the prediction results in a straight orange line that aligns with the maximum data points.

The second trial appears to yield a better prediction with a train loss of 0.0034763. Using a sample of the full dataset (data from 2015–2017) as the training set, the model predicts the same outcome but with a clearer view of the trend compared to the first trial: the number of Bitcoin transactions will increase in 2018, but it will experience a sudden dip in the middle of the year (Figures 1.1 and 1.2).

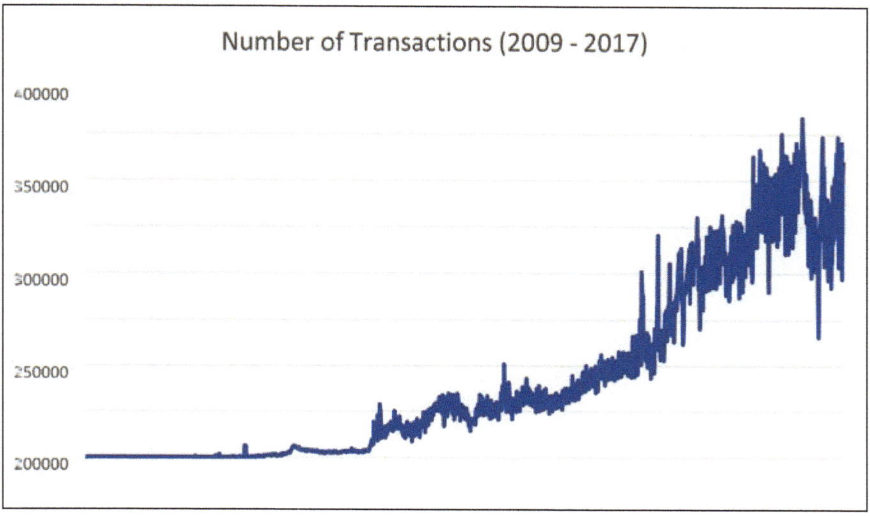

FIGURE 1.1 Full dataset for number of bitcoin transactions in period 2009–2017.

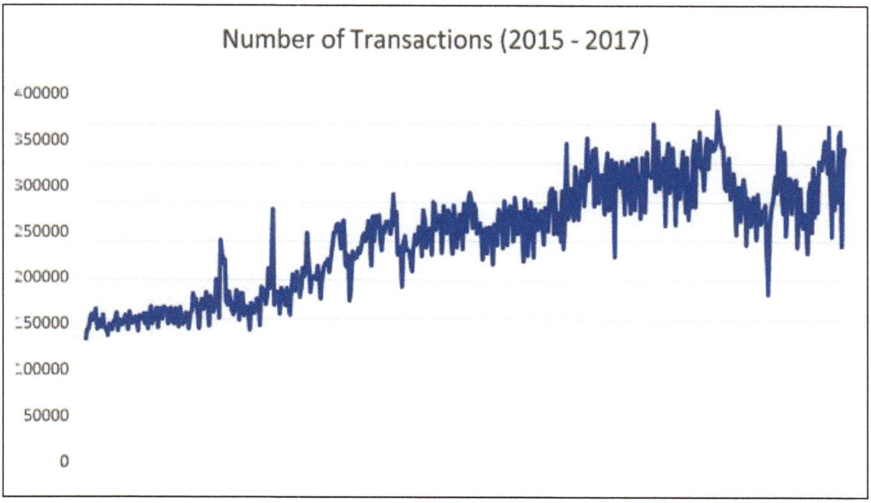

FIGURE 1.2 Sample dataset for the number of bitcoin transactions in period 2015–2017.

1.4 RESULT INTERPRETATION AND LIMITATIONS

This chapter is still lacking due to the limitation of Bitcoin's dataset (only 8 years of data) and the simple model of the neural network used to make predictions. Due to the different input datasets, the model gives two considerably different predictions on the exact number of Bitcoin transactions in the year of 2018. However, a predicted trend persists through both trials: the overall number of Bitcoin transactions will increase, but it is likely to experience a dip in amount in the middle of the year. This dip is the direct result of the volatility in the infancy period of Bitcoin as a currency. However, as many times before, the market will once again put its trust in Bitcoin's future as Bitcoin's popularity continues to increase over time.

The result of the first trial are shown in Table 1.3 and Figure 1.3; and the second trial results are shown in Table 1.4 and Figure 1.4.

1.5 CONCLUSION

Due to the infancy of Bitcoin, its long-term future is still shrouded in mystery. Nevertheless, the chapter has discussed at length the potentials (advantages and shortcomings) of Bitcoin in its development to become the global currency. Currently, these aforementioned potentials are mainly theoretically based, but in the short run, Bitcoin's trend to gain popularity and market share should be more apparent through the results of the neural network predictions.

KEYWORDS

- **bitcoin**
- **cryptocurrency**
- **financial market**
- **global currency**
- **globalization**
- **gold**
- **neural network model**
- **peer-to-peer electronic cash system**

TABLE 1.3 The First Trial Results

Optimization Finished!		***Phthon 3.6.3 Shell***		
Train loss: 0.023035				
[160241.546875	339506.28125	367355.625	367364.46875	367570.34375
367702.53125	367710.	367662.25	353990.3125	367563.09375
366303.90625	366043.1875	367708.90625	367615.21875	367698.03125
301127.40625	367701.09375	367709.65625	367400.90625	367373.40625
367709.71875	367445.46875	364404.34375	367710.	367700.875
367704.90625	367710.	367709.65625	367701.1875	367709.8125
367710.	367709.8125	367708.15625	367707.4375	367710.
367708.9375	367710.	367709.9375	362403.46875	367709.9375
367708.84375	367442.65625	367685.9375	367398.9375	366889.59375
348764.5625	367710.	363287.15625	367710.	365667.5625
367660.8125	367709.9375	367705.53125	367614.96875	367709.9375
367709.9375	276160.5625	367614.0625	367709.90625	367704.53125
367684.6875	367650.28125	367710.	367674.21875	367709.9375
367710.	367707.1875	349342.71875	367709.65625	367710.
328295.0625	367710.	367710.	367710.	367709.9375
367708.5	367710.	367709.15625	367188.4375	367680.21875
367710.	367693.	367709.9375	365730.3125	367709.9375
367710.	367701.1875	367710.	367693.90625	367710.
367710.	367709.8125	367710.	367710.	367710.

TABLE 1.3 *(Continued)*

Optimization Finished!		***Phthon 3.6.3 Shell***		
Train loss: 0.023035				
367702.84375	367710.	367710.	364609.875	367710.
367710.	367709.6875	367708.9375	367697.59375	367710.
367709.71875	367701.96875	367710.	365841.25	365341.875
367710.	367710.	367310.5	367710.	367710.
367710.	367680.3125	367710.	367710.	367710.
367709.5	367710.	367707.375	367710.	367710.
367710.4375	367709.15625	367691.625	367709.59375	364232.59375
367698.	367710.	367710.	367710.	364700.25
367710.	367710.	367710.	367710.	367710.
367709.90625	365678.875	367698.15625	367710.	367708.375
367709.875	367709.875	367709.71875	367707.5	367708.78125
367710.	367697.84375	367709.34375	367709.9375	367710.
367709.71875	367710.	367676.71875	367709.25	357290.78125
367709.9375	367709.28125	367709.90625	367701.1875	767709.9375
367630.6875	367709.90625	367698.46875	367652.9375	336067.3125
367710.	367709.9375	365439.6875	367709.875	367710.
367471.71875	367708.5	367709.59375	367709.9375	367709.5625
367709.875	367701.75	367709.78125]		

TABLE 1.4 The Second Trial Results

Optimization Finished!				
Train loss: 0.00347626				
[278033.125	328643.40625	329561.	323559.46875	
284486. 40625	334015.5	311113.875	302130.53125	350194.5625
348237.71875	356281.1875	315193.8125	340055.90625	
329066.09375	222709.40625	308345.96875	296495.15625	329293.1875
307134.15625	344862.25	342066.8125	327287.5625	352078.25
278517.84375	345262.375	275424.78125	272488.96875	328326.875
247429.375	322878.4375	334474.03125	320745.53125	
270540.90625	316062.9375	326024.0625	278113.25	346578.375
283435.8125	348955.3125	259986.5625	265882.5	317026.375
231175.65625	280376.875	309447.09375	336428.375	301726.59375
352480.96875	353879.28125	303342.15625	347860.25	322429.6875
348307.3125	275808.8125	347413.6875	358659.6875	341802.96875
353065.0625	345329.125	355281.4375	290082.375	327677.15625
350510.34375	323626.03125	350255.78125	334585.34375	354220.3125
282001.375	318983.3125	331636.3125	264889.71875	292331.3125
339513.9375	348161.65625	274338.28125	349178.0625	341065.21875
268503.3125	272987.03125	241915.90625	339887.	275836.09375
352287.53125	351340.96875	330999. 15625	350788.78125	
317725.28125	355444.125	233714.53125	356658.4375	361700.59375

TABLE 1.4 *(Continued)*

Optimization Finished!				
Train loss: 0.00347626				
343512.84375	360815.78125	346508.21875	363707.875	312253.59375
355022.	362046.46875	356908.6875	363681.03125	365330.5
366476.625	336288.9375	353939.4375	344613.5625	284506.90625
334295.625	326758.6875	357324.78125	308150.03125	354968.4375
323079.15625	303627.59375	258180.421875	219302.296875	
316916.78125	115374.046875	270198.125	182167.640625	
108152.7734375	208349.140625	104250.046875	202424.734375	
119823.7890625	239466.921875	161660.265625	183010.296875	
252439.265625	116055.796875	243070.90625	112856.0703125	
200945.90625	121604.078125	105414. 640625	231658.96875	178218.625
318348.59375	264057.34375	309282.09375	302088.125	323880.125
305711.53125	183116.	284734.5	125679.375	225260.234375
177788.09375	157983.296875	256057.796875	230998.9375	307104.09375
186503.984375	330929.25	241798.25	250735.96875	267358.8125
203566.546875	245699.40625	95442.859375	118939.34375	
111626.4609375	123484.3984375	262981.09375	193171.921875	
304688. 40625	257306.859375	240041.421875	187755.234375	
203776.578125	186650.09375	189822.484375	239317.703125	
40659.328125	291486.15625	313922.46875	272464.8125	311270.53125

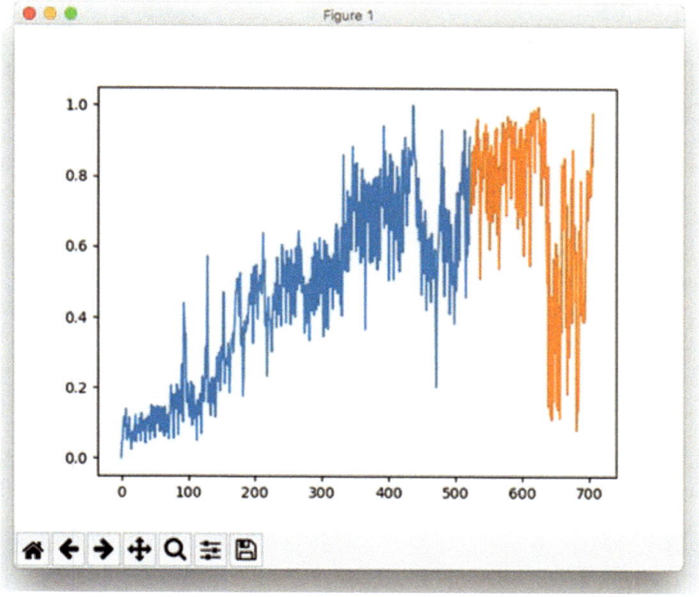

FIGURE 1.3 The first trial results.

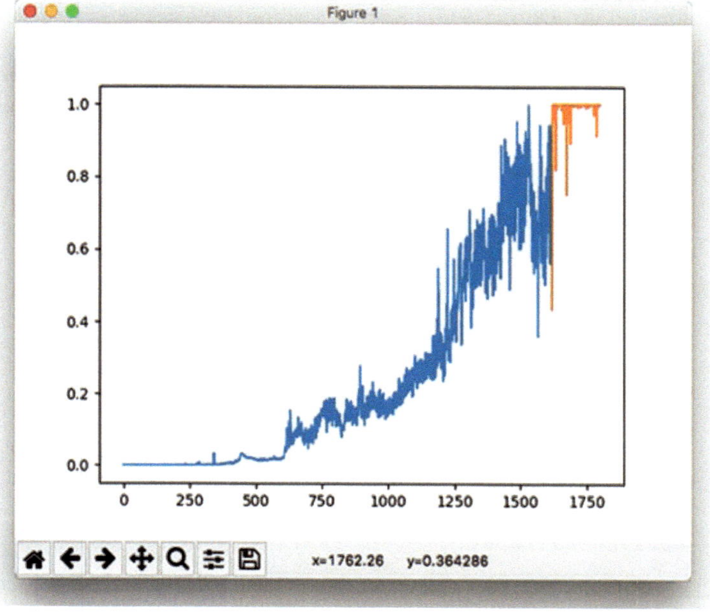

FIGURE 1.4 The second trial results.

REFERENCES

Bitcoin History: The Complete History of Bitcoin, [Timeline]. (n.d.). Retrieved from: http://historyofbitcoin.org/ (accessed on 11 January 2023).

Bouri, E., Jalkh, N., Molnar, P., & Roubaud, D., (2017). Bitcoin for energy commodities before and after the December 2013 crash: Diversifier, hedge, or safe haven? *Applied Economics, 49*(50), 5063–5073. doi: 10.1080/00036846.2017.1299102.

Buterin, V., (2012). *Anniversary of the Great Bubble of 2011*. Retrieved from: https://bitcoinmagazine.com/articles/anniversary-of-the-great-bubble-of-2011-1339139269/ (accessed on 11 January 2023).

Cawrey, D., (2013). *Could Deflation Cause Problems for Bitcoin?* Retrieved from: https://www.coindesk.com/deflation-and-bitcoins/ (accessed on 11 January 2023).

Chen, A., (2017). *We Need to Know Who Satoshi Nakamoto Is*. Retrieved from: https://www.newyorker.com/business/currency/we-need-to-know-who-satoshi-nakamoto-is (accessed on 11 January 2023).

Damien, A., (2017). *Example of Neural Network*. Retrieved from: GitHub: https://github.com/aymericdamien/TensorFlow-Examples/blob/master/notebooks/3_NeuralNetworks/neural_network_raw.ipynb (accessed on 11 January 2023).

Ermakova, T., Fabian, B., Baumann, A., Izmailov, M., & Krasnova, H., (2017). Bitcoin: Drivers and impediments. *SSRN Electronic Journal*. doi: 10.2139/ssrn.3017190.

Nakamoto, S., (2008). *Bitcoin: A Peer-to-Peer Electronic Cash System*. Retrieved from: http://nakamotoinstitute.org/bitcoin/ (accessed on 11 January 2023).

Number of Confirmed Bitcoin Transactions Daily. Retrieved from: https://blockchain.info/charts/n-transactions (accessed on 11 January 2023).

The History of the Euro, (2017). Retrieved from: https://european-union.europa.eu/principles-countries-history/history-eu_en (accessed on 11 January 2023).

What If the Bitcoin Bubble Bursts? (2017). Retrieved from: https://www.economist.com/news/leaders/21722841-latest-frenzy-tulipmania-gold-rush-or-dotcom-boom-what-if-bitcoin-bubble (accessed on 11 January 2023).

CHAPTER 2

Industry 5.0: A Solution for Leading SMEs

MONIKA ARORA,[1] PREMA MAHALE,[2] and KULDEEP BHALERAO[3]

[1]*Amity Business School, Amity University, Manesar, Haryana, India*

[2]*Vivekanand Education Society's Institute of Management Studies and Research, Chembur, Mumbai, Maharashtra, India*

[3]*Bharati Vidyapeeth's Institute of Management Studies and Research, Navi Mumbai, Maharashtra, India*

ABSTRACT

Small and medium-sized businesses (SMEs) are one of the main drivers of any country's economy. The strength of SMEs reflects the health of the country's economy in both developing and developed nations. Although the adoption of technological platforms is still in its early stages, it is true that they can help SMEs grow and develop their businesses to a new level. These businesses significantly contribute to GDP growth and maintain appropriate money mobility throughout the economy, according to studies on their impact on the growth of a nation's economy. In truth, governments depend on SMEs to advance their economies in a number of European and American nations, as well as Asian and African nations.

Numerous studies have been conducted on SMEs before, during, and after COVID-19 to determine the techniques used and highlight the advantages of adopting technology in terms of business performance. This research chapter utilized secondary data from various sources such

Sustainable Strategic Business Infrastructure Development and Contemporary Digital Practices in Industry 5.0. Indrajit Ghosal, Bikram Prasad, Nilanjan Ray, & Lalit Kumar Joshi (Eds.)

as research papers, case study materials, books, and reports. A qualitative analysis of the data was conducted for the study.

The usage of technologies has benefited SMEs, leading to improved performance and business sustainability. The study contributed to the development of a sustainable strategy for SMEs, outlining three goals: (i) researching the role of SMEs and Industry 5.0 in economic advancement, (ii) investigating the opportunities and challenges of implementing Industry 5.0 in SMEs, and (iii) proposing the Industry 5.0 model for SMEs.

The model with Industry 5.0 integration was proposed after the completion of an empirical investigation. This study makes a significant contribution to our current understanding of sustainability. The purpose of the study was to identify the strategies employed by SMEs. SMEs have embraced technology in various areas, including social media, mobile technology, ERP, cloud technology, artificial intelligence (AI), and Industry 5.0. The utilization of technology will also bring benefits to the economy and the growth of organizations in terms of productivity, revenue generation, satisfaction, and improved stakeholder communication and transparency.

A solid conclusion can be drawn for business owners, students, academics, managers, and SME policy makers. Furthermore, it will contribute to the development and enhancement of the country. Additionally, this will contribute to Industry 5.0 and provide advantages and opportunities on a global scale. The proposed paradigm can be further expanded upon and applied to the implementation of operational excellence in SMEs.

2.1 INTRODUCTION

According to the results of the study (Parida, Sjödin, & Reim, 2019), industrial processes have changed over the past 10 years, particularly in the field of digitalization. According to Zhou's (2013) study, information and communication technology tools are applied at every level of the production process (ICT). This causes a variety of issues with technology, logistics, organization, the environment, among other areas. A challenging transition process is currently underway and needs to be carefully managed. The usage of new technology has a tremendous impact on people's daily life and work (Ciffolilli & Muscio, 2018). People are involved in the process of altering industrial paradigms, whether they are participating in the sector as customers or as employees (Saniuk, Grabowska, & Gajdzik, 2020). To make a significant contribution, both customers and employees must be eager to acquire new skills, share their knowledge, and be adaptive (Lepore, Micozzi, & Spigarelli, 2021).

Malesios et al. (2018) noted that the adoption of the former may affect SME performance, demonstrating a correlation between various sustainability metrics and the two economic indicators. There is a strong positive correlation between specific sustainability characteristics and business success. The study's conclusions imply that a combination of technological advancement and economic expansion will enhance societal well-being and business productivity. Government initiatives, financial support for small businesses, and capacity building for human resources all contribute to the development of SMEs with a determination coefficient of 97.6%. It is suggested that the government implement a technology innovation-based economic growth plan in Makassar City, Indonesia, to increase the productivity of community economic companies there (Surya et al., 2021).

Industry 5.0 recognizes the capacity of the private sector to support social goals beyond employment and economic development. By ensuring that production respects the boundaries of our planet and prioritizes the health and safety of industrial workers, it aims to transform the private sector into a resilient producer of prosperity. Industry 5.0 complements the existing Industry 4.0 paradigm and will assist the European industry in transitioning to one that is resilient, sustainable, and focused on meeting the needs of people. Industry 5.0 is a clear result of the European Commission's recognition that social and environmental issues in Europe require better integration into technological innovation, and that a systematic strategy, rather than a focus on individual technologies, is necessary (Breque et al., 2021).

Supply chain management, cloud manufacturing, and intelligent healthcare have all been examined as potential applications for Industry 5.0. The authors then discuss a few enabling technologies for Industry 5.0, including edge computing, digital twins, collaborative robots (CoBot), the Internet of Everything, blockchain, and 6G and beyond networks (Maddikunta et al., 2022).

The "industrial revolution" is being driven by technological advancements that are fundamentally altering how business is conducted and impacting both society and the economy. While some of these changes are intentional and desired, others are unintended and undesirable. Similar to its predecessors, Industry 4.0 is driven by technology. On the other hand, Industry 5.0 is focused on values. Both Industry 4.0 and Industry 5.0 are necessary for technological breakthroughs and solutions. Industry

4.0 relies on Industry 5.0 to serve as a reminder of the essential social demands, values, and obligations as ultimate goals.

The research was conducted with the following objectives in mind: to study the role of SMEs and Industry 5.0 in economic growth, to investigate the potential and obstacles associated with implementing Industry 5.0 in SMEs, and to utilize Industry 5.0 for SMEs in the development of the proposed model.

The chapter discusses the usage of Industry 5.0 in SMEs, along with its prospects and problems. It also covers the employment of technology in areas such as intelligent healthcare, cloud manufacturing, and supply chain management. The next section of the literature review reviews the different functions that utilize intelligent, automated, and renewable resources. The concepts of Industry 5.0, government 5.0, and society 5.0 are used to further explore the subject. Additionally, a model will be suggested for SMEs to follow, which can be applied to any business.

2.2 LITERATURE REVIEW

According to the author, Industry 5.0 promotes sustainable development values through 16 functions. Examples of the suggested functionalities include supply chain flexibility, open sustainable innovation, renewable integration, intelligent automation, and circular intelligent products. These roles are closely tied to one another and should be developed in a precise sequence to optimize the value gains from sustainable development. The roadmap to Industry 5.0-driven sustainability developed in this study aims to provide readers with a better understanding of how Industry 5.0 can aid in sustainable development by outlining the best ways to manage the growth of its functions to maximize their synergies and contribution to the desired sustainability values. The chapter also identifies key directions for further investigation, emphasizing the elements that may support the growth of Industry 5.0, such as Corporate Governance 5.0 and Government 5.0.

Following is a summary of the three primary sustainability objectives of Industry 5.0 and their important contributions to many facets of sustainable development:

- This employs economic, supply-chain, and business resilience.
- Value network integration, intelligent automation, sustainable thinking, renewable integration, circular intelligent products,

operational and resource efficiency are the primary factors that support the goals of environmental sustainability.
- System integration and interoperability, data sharing and transparency, open sustainable innovation, upskilling and reskilling, employee technical help, and service orientation and personalization are the major drivers of Industry 5.0's human-centricity goals.

The findings highlight that the inclusive and sequential development of all sustainability functions is necessary for Industry 5.0 to make the desired contribution to sustainable development. Such an inference can be made for two main reasons. First, the findings indicate that each function may address only a few components of sustainable development. To offer a wide range of micro- and macro-economic, environmental, and social sustainability consequences, functions do indeed complement one another. Second, Industry 5.0's sustainable development functions have intricate linkages of precedence, and the order in which they developed would determine how Industry 5.0 would result in sustainable development values. Real-time communication (RTC) supports the growth of supply chain adaptability (SCA), demonstrating how SCA develops on the foundation of supply chainwide real-time communication. The elimination of functional silos throughout the supply network or the data-driven forecasting of impending disruptions are examples of the supportive roles played by RTC for SCA. Although RTC does not directly impact supply chain resilience, it significantly improves the adaptability of supply chains, which has significant indirect effects. Therefore, the collective and sequential growth of the identified Industry 5.0 functions is essential for sustainable development:

- The business, supply chain, and economic aspects of resilience are facilitated by features such as real-time communication, value network integration, intelligent automation, supply chain modularity, business process monitoring, service-orientation, personalization, operational and resource efficiency, and supply chain adaptability.
- Therefore, the collective and sequential growth of the identified Industry 5.0 functions is essential for sustainable development in the business, supply chain, and economic aspects of resilience are facilitated by features such as real-time communication, value network integration, intelligent automation, supply chain modularity, business process monitoring, service orientation, personalization, operational and resource efficiency, and supply chain adaptability (Ghobakhloo et al., 2022).

The future for SMEs is indeed bright, but it is also quite challenging. Future studies should focus on internal obstacles, such as conflicting business strategies, which may hinder the adoption of new technologies. Furthermore, researchers should explore whether the government should intervene in SMEs' financial issues or if SMEs should seek alternative sources of funding for their projects.

Another intriguing study could explore whether small and medium-sized enterprises (SMEs) can afford robotics as part of Industry 5.0, especially in the initial stages when new technology tends to be expensive. If they are unable to do so, it could result in a loss of competitiveness. The question then arises: can SMEs maintain their competitiveness long enough to survive until this new technology becomes more affordable?

SMEs also have a stronger focus on people, as a significant percentage of them are family businesses. In this context, one question that arises is whether SMEs will be able to culturally integrate the use of robots to assist humans (Andreas Karaoulanis, 2022, Vol. 26, No. 3S).

- The human-centric approach, which prioritizes the requirements of the worker and asks what technology can do for them, places the worker's needs at the center of the production process.
- Sustainability emphasizes waste reduction, resource conservation, and recycling of natural resources.
- Resilience, which denotes a move towards stronger industrial production, provides support through adaptable manufacturing capacity and flexible processes, particularly when a crisis arises.

The primary forces behind Industry 5.0 are:

- Human centricity addresses operator 4.0 and the Human Factor.
- In Industry 4.0, resilience manages technology consumption.
- Sustainability covers the industry's factors and sustainable advantages.

Based on the study, the use and implementation of Industry 5.0 in SMEs has been started.

2.3 ANALYSIS AND MODELING

There is a comparative study between Industry 4.0, Industry 5.0, and Society 5.0. This distinctive study has been conducted by multiple authors and presented by Demir (2019).

	Industry 4.0	Industry 5.0	Society 5.0
Motto	Smart factory	Bioeconomy Human–robot-Coworking	Anyone gets opportunities anytime, anywhere.
Motivation	Mass production	Sustainability Smart society	Humankind lives in harmony with nature.
Power source	Electrical power. Fossil based fuel. Renewable power sources.	Electric power. Renewable power sources.	Electric power. Renewable power sources.
Involved technologies	Internet of things (IoT). Cloud computing. Big data. Robotics and artificial intelligence (AI).	Sustainable agricultural production. Bionics. Renewable resources. Human robot coordination. Human robot collaboration.	Internet of things (IoT). Cloud computing. Big data. Robotics and artificial intelligence (AI). Virtual reality (VR). Human robot coordination. Human robot collaboration.
Involved research areas	Organizational research. Process innovation and improvement. Business administration.	Agriculture. Biology. Waste prevention. Organizational research. Smart environment. Process improvement and innovative business administration.	Smart environment. Process improvement and innovative business administration.

Source: Adapted from Demir (2019)

Industry 5.0 aims to leverage both human and machine capabilities, harnessing human creativity and power to enhance process effectiveness and efficiency. It involves a comprehensive examination of human intentions, encompassing image processing, audio analysis, and other sensory analyses. The goal is to establish a framework that combines human and machine cognition, enabling a holistic approach to work, encompassing 360-degree perspectives, audio analysis, and other sensor analyses (Sharma et al., 2022). The system will encompass all the necessary data for analysis.

The networking sector, data interoperability, multiscale, dynamic modeling, etc., play a crucial role in Industry 5.0. Digital drivers, shop floor trackers, virtual training, advances in sensor technologies, and machine cognition all heavily rely on simulations. Due to the rapid pace of technological breakthroughs, Industry 5.0 facilitates changes in regular corporate operations and the integration of human processes.

These primary forces will aid in the process of identifying stages and their effects on the development of the suggested model (Saptaningtyas et al., 2020). The proposed Industry 5.0 model can be summarized as follows with the help of Industry 5.0:

- **Stage 1: Identification and Analysis:** The current business process is examined, considering the database and its extensive utilization. The data structure and knowledge management are employed throughout the entire process. Unstructured data will be utilized, and the study can incorporate artificial intelligence (AI) and business data analysis.
- **Stage 2: Agile-based Strategy:** Organizations were created following this strategy, with a greater focus on human resources. The essential key aspects of this stage are fully utilizing product demand, regulations, production, product technology, capital, and human resources.
- **Stage 3: Product Design:** The product is designed with the client in mind. It will be entirely client-driven. The product cycle will take care of the product's consumers and thoroughly investigate the parameters in accordance with customer needs.
- **Stage 4: Product Innovation:** The product will be thoroughly tested during the design phase. Engineering for materials and processes will be included. Technology components such as data analysis, artificial intelligence, and virtual reality will also be incorporated.
- **Stage 5: New Business Model:** This involves product invention and mass manufacturing. In order to foster product innovation, the organization has implemented new business processes.

Every stage is connected to the one before it so that the quality can be checked, remedial action can be taken immediately, waste can be eliminated, and the system can be made more sustainable. Model suggested using Industry 5.0 is illustrated in Figure 2.1.

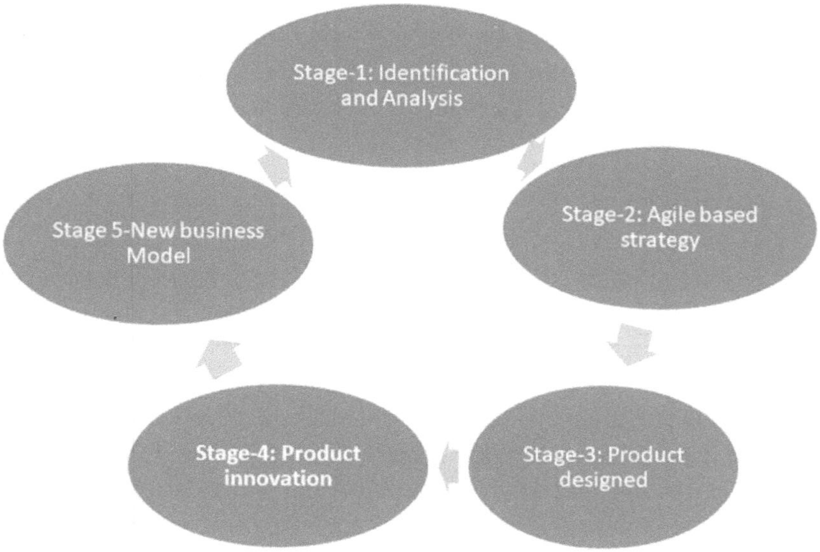

FIGURE 2.1 Proposed model using Industry 5.0.

Source: Authors.

Every stage is linked to the previous stage to allow for quality checks and corrective measures to be taken on-site. This helps in eliminating wastages and making the system more sustainable.

2.4 APPLICATIONS

There are many applications in SMEs, and the usage of Industry 5.0 has provided businesses with high-quality solutions. The following are some of the applications that were covered:

1. **Food Manufacturing:** It has been more efficient and successful thanks to product research, innovation, and technological utilization. Numerous compatible products have been developed as a result of client demand and requirement research.
2. **Smart Logistics:** It has utilized technology to acquire real-time warehouse inventory. Real-time tracking has enhanced operational and procedural efficiency while significantly reducing waste.

3. **Pharmaceutical Manufacturing Sector:** Jafari et al. (2022) encompasses the production of pharmaceuticals and related medical equipment. Furthermore, it promotes the utilization of robotics in various operations, enhancing their effectiveness and efficiency.

2.5 CONCLUSION

Participation in Industry 5.0 will provide the research with a new perspective. It has not only offered a solution to the problem, but also introduced innovative and original approaches to work. Industry 4.0 and Industry 5.0 have revolutionized the utilization of technology in SMEs. In order to adapt to the current advancements, SMEs must effectively execute their plans. In Industry 5.0, incorporating the human "touch" is one method to enhance manufacturing quality. Nowadays, extensive data, state-of-the-art robotic technology, and human beings are integrated with humanistic or human values.

There may be competition from goods made in other nations. In terms of national culture and aristocracy, they essentially enjoy a number of advantages. The products will still have distinctive qualities that will set them apart from those made by robots in Industry 5.0, even though these features are included in the manufacturing process. Industry 5.0 adoption will be necessary for SMEs to contribute to the success of their businesses.

KEYWORDS

- **cutting-edge robotic technology**
- **Industry 5.0**
- **pharmaceutical manufacturing sector**
- **small and medium-sized businesses**
- **SME performance**
- **SME sustainability**
- **technology adoption drivers**
- **technology benefits**

REFERENCES

Andreas, K., (2022). Kaunas university of applied sciences. Can small business survive Industry 4.0 and 5.0? *International Journal of Entrepreneurship, 26*(3S), 1–11.

Breque, M., De Nul, L., & Petridis, A., (2021). *Industry 5.0: Towards a Sustainable, Human-Centric, and Resilient European Industry.* Luxembourg, LU: European Commission, Directorate-General for Research, and Innovation.

Ciffolilli, A., & Muscio, A., (2018). Industry 4.0: National and regional comparative advantages in key enabling technologies. *European Planning Studies*, 2323–2343.

Demir, K. A., & Cicibas, H., (2017). Industry 5.0 and a critique of Industry 4.0. In: *Proceedings of the 4th International Management Information Systems Conference* (pp. 17–20). Istanbul, Turkey.

Demir, K. A., Döven, G., & Sezen, B., (2019). Industry 5.0 and human–robot co-working. *Procedia Computer Science, 158*, 688–695.

Jafari, N., Azarian, M., & Yu, H., (2022). Moving from Industry 4.0 to Industry 5.0: What are the implications for smart logistics? *Logistics, 6*(2), 26.

Lepore, D., Micozzi, A., & Spigarelli, F., (2021). Industry 4.0 accelerating sustainable manufacturing in the COVID-19 era: Assessing the readiness and responsiveness of Italian Regions. *Sustainability*, 2670.

Maddikunta, P. K. R., Pham, Q. V., Prabadevi, B., Deepa, N., Dev, K., Gadekallu, T. R., & Liyanage, M., (2022). Industry 5.0: A survey on enabling technologies and potential applications. *Journal of Industrial Information Integration, 26*, 100257.

Malesios, C., Skouloudis, A., Dey, P. K., Abdelaziz, F. B., Kantartzis, A., & Evangelinos, K., (2018). Impact of small- and medium-sized enterprises sustainability practices and performance on economic growth from a managerial perspective: Modeling considerations and empirical analysis results. *Business Strategy and the Environment*, 960–972.

Morteza, G., Mohammad, I., Muhammad, F. M., & Mobashar, M., (2022). Identifying Industry 5.0 contributions to sustainable development: A strategy roadmap for delivering sustainability values Morteza. *Sustainable Production and Consumption*, 716–737.

Parida, V., Sjödin, D., & Reim, W., (2019). Reviewing literature on digitalization, business model innovation, and sustainable industry: Past achievements and future promises. *Sustainability*, 391.

Saniuk, S., Grabowska, S., & Gajdzik, B., (2020). Social expectations and market changes in the context of developing the Industry 4.0 concept. *Sustainability*, 1362.

Saptaningtyas, W. W. E., & Rahayu, D. K., (2020). A proposed model for food manufacturing in SMEs: Facing Industry 5.0. In: *Proceedings of the International Conference on Industrial Engineering and Operations Management* (pp. 10–12). Dubai, UAE.

Sharma, M., Sehrawat, R., Luthra, S., Daim, T., & Bakry, D., (2022). Moving towards Industry 5.0 in the pharmaceutical manufacturing sector: Challenges and solutions for Germany. *IEEE Transactions on Engineering Management.*

Surya, B., Menne, F., Sabhan, H., Suriani, S., Abubakar, H., & Idris, M., (2021). Economic growth, increasing productivity of SMEs, and open innovation. *Journal of Open Innovation: Technology, Market, and Complexity*, 20.

Zhou, J., (2013). Digitalization and intelligentization of manufacturing industry. *Advances in Manufacturing*, 1–7.

CHAPTER 3

Strategies for Channelizing Women Artisans as Entrepreneurs to Leverage the Indian Economy: An Empirical Study of MSME (Handicraft Sector) Concerning ODOP in Uttar Pradesh

UMA SHANKAR YADAV, RAVINDRA TRIPATHI, and MANO ASHISH TRIPATHI

Department of Humanities and Social Sciences, Motilal Nehru National Institute of Technology, Prayagraj (Allahabad), Uttar Pradesh, India

ABSTRACT

Women who recognize opportunities, effectively manage resources, embrace new ventures, and take calculated risks to achieve their goals in establishing innovative companies are the leaders of a sustainable economy. The COVID-19 pandemic had a profound impact on the entire world, with India being one of the most affected nations. Many returning immigrants found themselves unemployed. However, women, particularly those in rural and semi-urban areas, gradually emerged as new business owners across various sectors. Microbusinesses thrived as the pandemic caused a decline in demand and supply. Notably, there was a high demand for handcrafted masks, Moonj crafts, bamboo crafts, and wooden handicrafts. Women entrepreneurs have proven their ability to overcome numerous challenges, including sustainability, achieving a balance, promoting

Sustainable Strategic Business Infrastructure Development and Contemporary Digital Practices in Industry 5.0. Indrajit Ghosal, Bikram Prasad, Nilanjan Ray, & Lalit Kumar Joshi (Eds.)

equitable development, navigating diverging economies, and maximizing commercial opportunities in today's competitive marketplace.

The implementation of the One District, One Product (ODOP) scheme in Uttar Pradesh in 2018 played a significant role. Despite the economic slowdown caused by the COVID-19 pandemic, the ODOP scheme gained popularity throughout Uttar Pradesh and contributed to boosting the state's economy. During this period, women and artisans acquired digital skills, collaborated with family members, and underwent a mindset shift towards entrepreneurship in Uttar Pradesh. Even though the One District One Product (ODOP) scheme was initially implemented in Japan in 1979, it has now gained immense popularity in Uttar Pradesh. Thanks to the nationwide campaign, ODOP is once again expanding. The Aatmanirbhar Bharat Abhiyaan and "Vocal for Local" initiatives, along with the development of women-focused handicraft companies such as wooden work, stone carving, toy making crafts, moonj craft, and bamboo craft (MSMEs), have greatly contributed to the popularity of ODOP. The purpose of this study is to assess the growth trends in women's entrepreneurship and the impact of post-pandemic micro-enterprises (handicrafts) on women. Additionally, it examines the types of organizations, the nature of activities, and the challenges faced by women entrepreneurs, as well as the entrepreneurial activities of women and women-led businesses. Finally, the opportunities provided by micro-entrepreneurial operations are explored.

3.1 INTRODUCTION

Enterprises run by women are continuously growing in many countries that collaborate to enhance income and strengthen national economies. However, women face significant challenges due to time constraints, social and physical restrictions, which hinder their ability to develop female-led businesses. To address these issues, it is crucial to provide female entrepreneurs with the necessary resources, tools, and support through both public and private sector entrepreneurial programs. These programs should focus on promoting best practices, addressing the specific needs and challenges faced by women entrepreneurs, and offering practical recommendations and guidance. Additionally, it is important to establish a centralized platform that serves as a clearinghouse of programs, data, and emerging research in order to further support female entrepreneurship. Female entrepreneurs have made a significant contribution to economic growth and poverty reduction worldwide. In countries like the USA, China, and Canada, female-run firms are growing at a rate more than twice that of other firms, contributing approximately $3 trillion to the country's economy and creating 23 million jobs. Moreover, in developing countries, the number of women-led micro and small businesses is also increasing, with an estimated 10 to 12 million women serving as job providers. Approximately 31% of businesses owned by women in the US are franchises or small businesses.

Entrepreneurial support for females is crucial in responding to the increasing demand for tools to integrate gender equality in both the private and non-government sectors. These support programs aim to promote development, address the specific needs of women entrepreneurs, and tackle the challenges they face. Two key functions have been designed: providing recommendations, practical guidance, and support, as well as serving as an emerging research clearinghouse for programs and data (Khurana et al., 2021). Approximately 31% of businesses owned by women in the US are franchises or small businesses:

- 17% of women in African Black women are busy running a new business.
- Women's entrepreneurial rate in the USA is 16.6%.
- Between the ages of 11 and 50, approximately 14% of women-owned businesses employ workers.
- Women in Latin America and Caribbean islands represent 50% of entrepreneurs.
- 36.1% of immigrants in Canada are women business owners.
- With India and Southeast Asia less than 20% of women entrepreneurs.
- Controlled enterprises and those owned and created by women provide direct employment for approximately 27 million workers in India.

The data has been provided by the World Bank Group, which has conducted research on female support and created a clearinghouse of programs for entrepreneurship. The research focuses on the situation of females in micro-handmade, small agribusiness, and handmade businesses (Alawat & Renu, 2018; Al-Dafri & Alasania, 2020). It is known that women entrepreneurs are generally unskilled and start new businesses in the unorganized and informal sectors, or in their traditional women sector businesses. Worldwide, in the non-agricultural labor force, the majority of females, about 30%, are self-employed in the traditional or handmade sector. In the continent of Africa, this value rises to 75%. Female-based businesses are particularly concentrated in rural areas and are often home-based. These sectors are small, interpreted as traditional sectors, and mostly include services and retail. These are some important factors that influence female entrepreneurs. However, female entrepreneurs in the unorganized sector outnumber those in the organized sector, and out of all female enterprises, 98% are micro-enterprises, 1.8% are in the small sector, and only 0.5% are medium enterprises (Al-Dafri & Alasania, 2020).

3.1.1 IMPACT OF THE PANDEMIC ON THE MARKETING OF THE HANDICRAFT INDUSTRY AND WOMEN ARTISANS

The COVID-19 pandemic has had a widespread impact on all sectors of the economy, including primary, secondary, and tertiary sectors. However, in India, women, children, and the elderly have been particularly affected by this crisis. The lockdown measures have caused significant damage to businesses and livelihoods. The handicraft sector, in particular, has been directly or indirectly affected by the pandemic. Despite this, women and local artisans have shown resilience by producing handmade masks, baskets, pottery, terracotta, woodcraft, and other products from their homes. However, the overall pace of these industries has slowed down due to the lack of demand in the market caused by the shutdown. Consequently, the marketing of these products has also been adversely affected. Nevertheless, women in the handmade industry have supported their male counterparts by producing masks and household products, thereby earning money and providing support to the community. Additionally, Indian women have embraced digital entrepreneurship during this time.

3.1.2 REGARDING HANDICRAFT PRODUCTS NEW SCHEME (ODOP) IN UTTAR PRADESH

Handmade products of Uttar Pradesh are covered under the ODOP program, which stands for One District One Product. This scheme was initially launched by the UP government and has received support from the Japanese government since 1979 (Chawla, 2013). The ODOP program aims to address job problems and prevent migration by providing local employment opportunities in Uttar Pradesh and across India (Sahoo, 2020; Yadav, Ravindra, & Mano, 2022i). It offers marketing platforms, both traditional and digital, for the 75 districts' special handicraft products. Marketplaces like Hunnar Haat, Saras Mela in Greater Noida, and Avadhsilp Gram provide marketing facilities for the handicraft sector, with many women selling their products in these markets. The scheme has the potential to empower artisans, skilled workers, especially women, and improve the socio-economic and financial situation of unorganized workers, rural communities, and tribal people in Uttar Pradesh. During the national shutdown of small industries, especially manufacturing, artisans faced significant challenges, including job unavailability and financial difficulties. It is now crucial to rethink and revive our traditional crafts, which are eco-friendly and create local-level markets, particularly for the handicraft sector in Uttar Pradesh, in line with the vision of Atmanirbhar Bharat and Vocal for Local (CII, 2020).

Khan et al. have focused on the significant problem faced by handicraft markets and how we can improve the quality of women artisans in Uttar Pradesh (John, 2017). They have observed that the handicraft sector is heavily reliant on the local market, fairs, consumers, local demands for these products, and how these products are connected to our culture and heritage. They have also discussed the "One Station One Product" scheme and the role of women. In February 2022, Finance Minister Nirmala Sitharaman announced that the scheme will be implemented at some famous Indian railway stations, showcasing the heritage and traditional crafts of their respective cities. Women will play a crucial role in the promotion and marketing of these handicraft products, as 60% of the stall opening certificates will be given to women artisans and entrepreneurs. This scheme aims to create a new market hub at Indian railway stations and popularize handicraft skills both nationally and internationally. It will attract tourists and travelers who are interested in learning about the origin of these crafts and the creative designs of Indian craftsmanship. This will provide additional benefits to artisans, especially women artisans from Uttar Pradesh. Terracotta craft stalls will be opened at Gorakhpur station, Moonj craft stalls at Prayagraj junction, leather products at Kanpur station, and chikankari craft at Lucknow junction.

3.2 REVIEW OF THE LITERATURE

The following literature suggests the goal of the article. Sahoo (2021) conducted research on the role of the 4Ps in the crafts sector. He observed that Indian artisans have the ability to create a wide range of patterns and proposed the four Ps: promotion, placement, price, and product. In a previous study (Anand & Bibha, 2020), it was emphasized that handmade products are produced with the assistance of Self-Help Groups (SHGs) and NGOs involved in this sector. These organizations receive financial support from the finance industry, as well as grants, subsidies, and funds from related ministries in the country and the state of Uttar Pradesh. The study also discussed the role of SHGs and other private organizations, known as NGOs, in the marketing and manufacturing of the Chikankaari craft of Lucknow.

The impact of pandemics on small businesses has been examined by GoI (2020). They propose that the government should provide financial assistance to the handicraft business. Ghosal & Prasad (2020b) and Kapoor (2018) emphasize the role of suppliers in the Indian craft business. GoI (2020) also highlights the importance of handcraft businesses and investigates the risks and challenges faced by small and female-based industries. The researcher employs descriptive, experimental, and analytical techniques in their empirical analysis. Jochhon & Indrajit (2019) and GoI (2020) examine the positive influence of handmade products such as craft, handloom, terracotta, woodcraft, textile, Black pottery, Jardozi, and education about stone carving and wooden craft in vocational education and skill training. The study paper emphasizes the dependence of the handicrafts-based industry on traditions and local customs. Kapoor (2018) and Jafari et al. (2020) discuss the significance of a cutting-edge and highly effective methodology in training and teaching about handmade products, as well as the recognition and preservation of their unique characteristics.

Academics, researchers, and students can comprehend the value and potential of handmade products in terms of production, marketing, and supply techniques. They can utilize their knowledge, intelligence, and skills to enhance the earnings of workers and artisans involved in the handicraft industries. These findings indicate a significant difference in the attitude and appreciation of crafts among learners in both the control and experimental groups (Magiya et al., 2011).

Khurana et al. (2021) have discussed an important objective in their article, focusing on the significance of Hindustani craftworks, the globalization of Indian crafts, and the arts of India's rich cultural heritage.

Magiya et al. (2011) conducted their research in the Bhadohi Moradabad, Meerut, and Banda districts of Uttar Pradesh. They reported an astonishing shortfall in revenue of over 70% in the handicraft sector, which is a part of the MSME sector, in the state. The handmade product industry experienced a complete collapse in working capital and cash flow in recent years, primarily due to the demonetization declared by the Union Government in 2016. Just as recovery was underway, the COVID-19 pandemic further exacerbated the situation (Khan & Amir, 2013).

Yadav et al. (2022c), in their project submitted to ICSSR, observed over 7,000 MSMEs, particularly handicraft enterprises, during the small industries shutdown. They found that 71% of these industries were unable to pay salaries to their workers or artisans due to the collapse of the marketing system caused by the lockdown declared in March 2020 (Yadav, Tripathi, & Tripathi, 2022a).

In their study, Yadav et al. (2022b) emphasized the necessity of establishing a robust digital institution and developing a national index to support artisans, enhance competition, and boost the export of handicrafts from Uttar Pradesh to other countries. They also highlighted the importance of providing a special package for the handicraft sector or including it in the PLI scheme by the government, enabling the sector to thrive and compete in the global handicraft market. The State Profile of the PhD Chamber of Commerce and Industry, Uttar Pradesh (2011), has proposed the Global Handicraft Index. It suggests possible parameters and solutions to improve the production and export of handmade products, as compared to machine-made products in the current scenario. Implementing these measures will enhance the marketing value of handicraft products. In the same year, he analyzed the initial analysis of the transformation and innovation of the handmade carpet industry, as well as the types of institutions and training centers that can enhance production and value. Yadav et al. (2022b) also emphasized that workers and suppliers in small industries or the handmade sector, such as gig workers, sell the products and send the money to the owners of the handicrafts before returning home. This traditional gig work differs from platforms like Amazon and Flipkart, which are generally assumed to be gig work by some researchers. Marketing for handicraft products and mediators is not similar to marketing for higher education, BPO, telecommunications, or Nike. Handcraft management experts focus on product marketing, while others emphasize the need for mediating and essential services (Yadav,

Tripathi, & Tripathi, 2020; Gitanjali & Nivedita, 2021). The concept of marketing was developed with the famous four Ps: place, promotion, product, and price. However, additional components are required for marketing beyond service marketing. These include partnership, packaging, position, programming, and people (Yassir Haleem, 2015).

3.3 OBJECTIVES

- To study the women artisans in the handicraft sector of Uttar Pradesh.
- To study the impact of a pandemic on the handmade industry, particularly on women entrepreneurs.
- To propose marketing and development strategies for the handicraft industry and provide suggestions on empowering women to become entrepreneurs in the digital era.

3.4 RESEARCH METHODOLOGY

A descriptive study has been conducted. For primary data collection, a structured questionnaire and interview schedule method were employed. The study respondents were selected from a Self-Help Group (SHG), a family, textile industry artisans, and crafts workers in Uttar Pradesh. Stratified and simple random sampling techniques were used in the current study. The data collection period spanned from March 2020 to January 2022. The study's sample size was determined to be 268 according to the Krejcie-Morgan rule, and it was drawn from a small number of districts in Uttar Pradesh, including Prayagraj, Bhadohi, Varanasi, Banda, and Moradabad. The statistical tests used in the study to evaluate the stated hypotheses were correlation and ANOVA. Secondary data was gathered from publications such as books, journals, and other sources. Information from the 74th cycle of the NSSO, NITI Aayog reports, Directorate MSME reports, and the Union budget for 2021 and 2022.

3.5 DATA ANALYSIS

Data has been examined using a stratified random sampling technique and relevant tools and software, namely SPSS.

According to Table 3.1, 16.7% of respondents are under the age of 30, 22.3% are between the ages of 30 and 40, and 57.3% are between the ages of 40 and 60. Around 2.9% of the responders are 60 years of age or older.

TABLE 3.1 Respondents' Ages

Response (Age of Participant)	F	Percentage (%)
Below 25–35	40	13.3
35–45	65	26.7
45–50	153	59.8
50	10	1.9
Total	**268**	**101.7**

Source: Authors' computations using original data.

Table 3.2 shows that 69.4% and 30.5% of respondents are male and female, respectively. This suggests that female participation is lower than male.

TABLE 3.2 Sex of the Participant

Response	Frequency	Percentage (%)
Men	186	66.5
Women	82	33.5
Rather than male female	0	0
Total	**268**	**100**

Source: Primary data computed by authors.

Table 3.3 shows the problems faced by handicraft workers during the COVID-19 pandemic. 25.3% of workers reported very low wages, 16% reported a lack of government support, 18.6% experienced personal financial problems, 4.11% faced poor working conditions, 2.6% dealt with a lack of adequate raw materials, 14.1% struggled with high living expenses, 22.1% faced a lack of employment, and 0.72% cited other reasons such as man-made and natural calamities, shifts in demand, and the emergence of foreign products, among others.

Table 3.4 describes that 3.3% of respondents have less than one year of experience, 4.47% of respondents have 2–3 years of experience, 38.4% of respondents have 4 to 7 years of experience, and 53.7% of respondents have 7 years or more of experience. This data was observed during the pandemic in 2020.

TABLE 3.3 The Problem Faced by the Respondents in the Pandemic Timetable

Response	F	Percentage (%)
Very low wages during COVID-19	68	23.3
No government support	43	18
Personal financial problem	50	16.6
Poor working condition	3	3.11
Raw materials	7	4.6
Living expenses are high	38	12.1
Lack of employment before the pandemic	60	22.1
Other reasons	2	0.72
Total	**268**	**100**

Source: Primary data and computed by authors.

TABLE 3.4 Experience of the Respondents During a Pandemic

Response	F	Percentage (%)
1 year	9	3.3
2–3 years	12	4.47
4–7 years	103	38.4
7 years and above	144	53.7
Total	**268**	**100**

Source: Primary data and computed by authors.

Table 3.5 shows that 4.8% of handicraft workers receive wages below Rs. 1,000. During and post-pandemic, 69% of handicraft workers receive wages between Rs. 1,001–2,000. Additionally, 18.6% of craft workers receive wages between Rs. 2,001–3,000, while 3.7% of handicraft workers receive wages between Rs. 3,001–4,000. Lastly, 3.7% of handicraft workers receive wages above Rs. 4,001 per week.

Tables 3.6 and 3.7 describe the satisfaction level of respondents in the handicraft sector. 18.2% of respondents agree with the satisfaction level in the working handicraft sector, while 16.2% of workers disagree with the satisfaction level in the handicraft sector. Additionally, 44.7% of wage workers are neutrally satisfied with working during the pandemic.

TABLE 3.5 Wages Per Week During Pandemic and Post-Pandemic for Participants

Response	F	Percentage (%)
Below Rs. 1,000	13	4.8
1,001 Rs. 2,000	185	69
2,001 Rs. 3,000	50	18.6
3,001 Rs. 4,000	10	3.7
Above Rs. 4,001	10	3.7
Total	**268**	**100**

Source: Primary data and computed by authors.

TABLE 3.6 Satisfaction Level of the Respondents in the Handicraft Sector in Post-Pandemic

Response	F	Percentage (%)
Agree	49	18.2
Disagree	43	16.2
Neutral	120	44.7
Strongly agree	27	8.9
Strongly disagree	29	10.8
Total	268	100

Source: Primary data and computed by authors.

TABLE 3.7 Satisfied with the Wage of the Respondents

Response	f	Percentage (%)
Agree	27	10.7
Disagree	74	27.6
Neutral	68	25.37
Strongly agree	58	21.6
Strongly disagree	41	15.29
Total	**268**	**100**

Source: Primary data and computed by authors.

Table 3.7 explains the workers' satisfaction level with wages. 27.6% of workers are satisfied with their wages, 10.7% of workers disagree with their wages, 25.3% of workers are usually satisfied with their wages, and 21.6% of workers strongly agree with the wages they earn. Additionally, 15.2% of workers strongly disagree with their wages.

3.5.1 TESTING OF HYPOTHESIS

- **H0:** There is no significant difference between the financial impact and age for handicraft workers during a global crisis.
- **H1:** There is a significant difference between the financial impact and age for handicraft workers during a global crisis.
  Table 3.8 shows the results of the Anova test. It is found that the p-value is 0.012. If this value is less than 0.05, the null hypothesis is rejected and the alternative hypothesis is accepted (Yadav et al., 2022h).
- **H2:** There is a significant difference between age and financial impact for handicraft workers.

TABLE 3.8 ANOVA

Source of Variance	Sum of Squares	Df	Mean Square	F	P-Value
Between en groups	2.465	3	0.822	109.6	0.012
Within n group	83.615	264	0.427	–	–
Ps Total	**86.08**	**267**			

Source: Primary data and computed by authors.

3.5.2 TESTING OF HYPOTHESIS

- **H0:** There is no significant difference between the impact on handicraft workers due to employee-related factors and age during a global shutdown.
- **H1:** There is a significant difference between the impact on handicraft workers during the pandemic time based on employee-related factors and age.
  The results of the analysis are presented in Table 3.9. The null hypothesis has been rejected, and the alternative hypothesis, "There

is a substantial difference between age and employee-related impact for handicraft businesses in the pandemic period," has been accepted. The p-value is found to be 0.000, which is less than the significance level of 0.05.

TABLE 3.9 Correlations

	Age	Employee Related Impact
Pearson correlation	1	0.445
AGE	–	0.000
Sig. (2-tailed)	–	
N	268	268
Pearson correlation	0.445	1
Employ EE relate d impact.	0.000	268
Sig. (2-tailed)		
N	268	

Source: Primary data and computed by authors with help of SPSS.

3.5.3 TESTING OF HYPOTHESIS

- **H0:** For handicraft business owners and employees during the post-pandemic, there is no discernible relationship between marketing influence and age.
- **H1:** For handicraft business owners and employees in the post-pandemic era, there is a significant age gap and marketing impact.

3.5.4 TESTING OF HYPOTHESIS

- **H0:** For handicraft artisans, there is no gender difference in the impact of exports.
- **H1:** There is a gender disparity in the effects of exports on hand-craft workers.

The findings of the correlation are displayed in Table 3.10. It has been determined that the p-value is 0.000, which is less than 0.05. Therefore, the alternative hypothesis, "There is a significant difference between gender and the impact of exporting on handicraft entrepreneurs during a pandemic," is accepted instead of the null hypothesis.

TABLE 3.10 ANOVA

Variance	Sum of Squares	Df	Mean Square	F	Value
Between en groups	0.046	1	0.046	0.196	0.000
Within groups	81.143	264	0.233		
Total	81.189	265			

Source: Primary data and computed by authors.

3.6 RECOMMENDATION AND STRATEGIES

The findings and strategies have been discussed in subsections. Approximately 27.6% of workers, who constitute a significant portion, are not satisfied with their current wages. These wages are earned from the handicraft industry (Yadav et al., 2022c).

3.6.1 STRATEGIES

3.6.1.1 TRANSFORMATION OF TRADITIONAL GIG WORKERS INTO DIGITAL GIG WORKERS

This gig terminology was introduced before 546 BCE during the Mauryan period in India and the Nanda period at that time. The marketing of handmade pottery, woodcraft, pot hacker basket sellers, and bowl sellers, known as "hackers" or "ferry walas" in the Indian language (Yadav, Tripathi, & Tripathi, 2022e), holds great significance. These traditional freelancers play a vital role in our society. Reviving this tradition is crucial as it not only generates employment opportunities but also garners a better response compared to machine-made products. Moreover, it remains accessible to the common people. In the era of digital marketing, artisans, especially women, are adapting and excelling in this field with the help of mobile apps such as Amazon, Flipkart, Snapdeal, and Meesho. These platforms enable them to sell their crafts, enhance their digital marketing skills, and even contribute to reducing the spread of corona viruses during the modern and challenging times of the pandemic (Tripathi, Tripathi, & Yadav, 2022).

3.6.1.2 E-MARKETING

The purpose of e-marketing is to utilize the internet as a source for small industries such as the handmade sector, cottage industry, handloom industry, and food processing industry incorporated in the ODOP scheme. This concept will help women artisans transform into entrepreneurs, as traditional marketing may not be suitable for them due to safety and security concerns in remote areas. However, in e-marketing, women can sell their handmade products through online bookings and receive payments via online payment modes. To facilitate this, there is a need for proper regulations and laws to promote systematic e-marketing for handmade products, as well as to support traditional marketing through fairs, Haats, local Melas, and religious Melas such as the Nauchandi Mela in Saharanpur, Avadhshilp Mela in Lucknow, and Kumbh Mela in the Prayagraj district of Uttar Pradesh (Yadav & Tripathi, 2022f).

3.6.1.3 HUNAR HAAT AND BRAND BUILDING

This is a common marketing scheme launched by the central government to promote handmade products in India and conserve traditional crafts in Uttar Pradesh, as well as other handmade products from Tamil Nadu, Rajasthan, Chhattisgarh, Madhya Pradesh, West Bengal, and the northeast region. The scheme aims to support minority artisans and entrepreneurs, enhancing their skills, livelihoods, and global awareness of their traditional values. These workers are also involved in food processing and freelance sales of pots and other crafts. The Hunnar Haat sources are exemplified by programs like SVEP (Start-Up Village Entrepreneurship Program).

3.6.1.4 INNOVATIVE INDIA AND DESIGN INDIA

This is especially for handmade craftsmen that innovate and make design. There is a need for the best cultural relations at the global level and even with the neighboring country and that nation which are rich in their traditional craft like Iran, Iraq Azerbaijan, and Russian Federation, there is a need for a soft cultural policy at place of hard defense policy among the countries for better export of handmade product, there should be organized cultural fare and trade fare in another country.

3.6.1.5 GIG WORKERS IN THE HANDICRAFT SECTOR A PIONEER IN MARKETING

I see that the term "gig" was introduced by a foreign expert during the advent of online marketing. However, marketing strategies in the handicraft sector have been in existence since ancient times. Artisans would travel from one village to another, showcasing and selling their handmade products, thereby generating income. This traditional approach attracted customers and promoted the handicraft market. In today's digital age, we can leverage the assistance of gig workers who can provide door-to-door or home delivery services for handicraft products. Additionally, women can also participate as gig workers in the near-home delivery sector (GoI, 2020):

- Handmade Products should be a global standard.
- There should be an organization of regular exhibitions in handicraft parks, museums, or any fair, where it should be mandatory to buy handmade products made by women in India.
- Consumer to consumer marketing of handmade products.
- Providing awards to gig workers who create handmade supplies instead of machine-made and electronic supplies. There should be a provision for 2 minutes of advertisements on TV after each show on media platforms such as YouTube, TV, movies, and other online shows and social media. These strategies will help increase the market demand for handmade products.
- Proper add on seaport about the traditional craft in which women should be brandabesdar. These strategies will help empower female entrepreneurs and self-help groups (SHGs) to compete with global NGOs involved in the handicraft industry.
- India should adopt a lenient cultural and trade policy for the export and import of handicraft products. Additionally, it is crucial to provide support and encouragement to women entrepreneurs in this sector to enhance the percentage of women's participation in the handicraft industry (Kapoor, 2018).
- Strategies for the development of the handicraft industry of Uttar Pradesh include the establishment of parks and handicraft marts in different districts. These include the carpet park in Bhadohi, Terracotta park in Gorakhpur, brass metal park in Moradabad, Pottery park in Khurja, Moonj park in Prayagraj, handloom park in Mirpur

and Subhadra, Chikankaari park in Lucknow, embroidery park in Badaaun, paper park in Saharanpur, and handicraft women park in Lucknow.

3.6.1.6 TECHNOLOGY WORKSHOP CENTER

A proper workshop should be organized to explore new and advanced technology in the field of handicraft work. Additionally, national seminars, workshops, and international conferences should be organized to promote the handicraft industry and encourage artisans to become entrepreneurs. It is also suggested to improve transport connectivity and recognize outstanding artisans through cash prizes or fellowships, similar to those awarded to students in the handicraft sector (Yadav, Tripathi, & Tripathi, 2022e). Please refer to Figure 3.1 for more details.

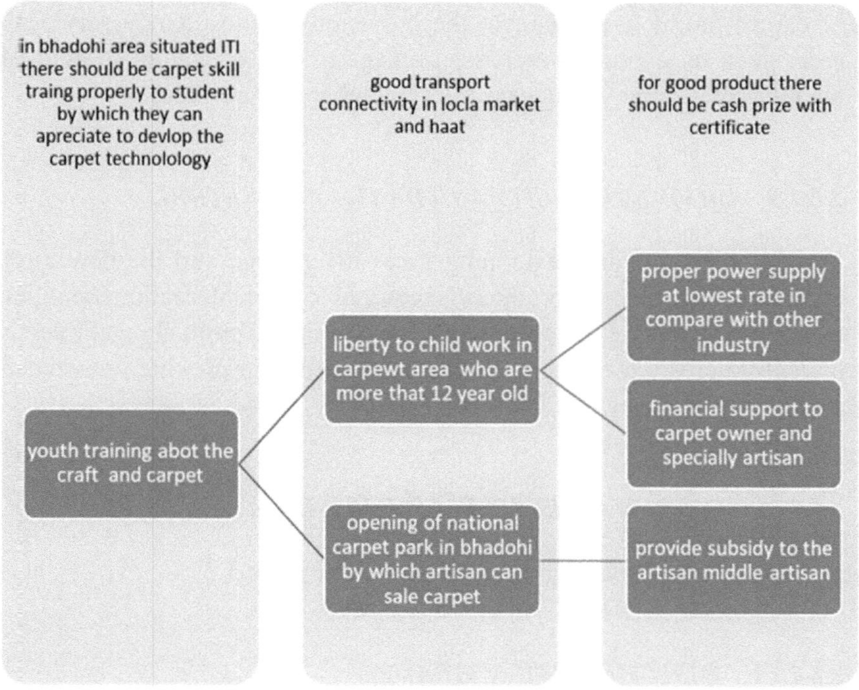

FIGURE 3.1 Technology transfer strategies for artisans' welfare.

Source: Figure has been designed and framed by the author's focusing strategies.

3.6.1.7 MARKETING STRATEGIES FOR WOMEN SHGS AND FACILITATION CENTER

In Varanasi, there is a handicraft trade facility center and a national level Rashtriya Hast Kala center that provide facilities to entrepreneurs, as well as a marketing resource center for the Indian handmade sector. The marketing center in Bhadohi and Varanasi also offers special facilities for women artisans and entrepreneurs. These centers were established in March 2019 for the development of a design center for handicrafts (Yadav et al., 2022g).

3.6.1.8 PROCEDURES/MANAGEMENT AND TRAINING PROGRAM AND WORKSHOP FOR PACKAGING AND EXPORT

To minimize the ratio of damage to products, it is necessary to have a well-equipped workshop and seek guidance from management experts in the marketing of handicraft products. This will enhance the marketability of export products and enable the provision of consultancy services in areas such as export awareness, packaging, and educating individuals, manufacturers, and young exporters (Shankar et al., 2022a).

3.6.1.9 DIGITAL PUBLICITY AND DIGITAL MARKETING

The organizations are developing innovative items, and the new digital marketing of handicraft products is required to give international publicity to explore the market for their items. Well-designed publicity and effective marketing create awareness among consumers and provide potential for producers (Yadav et al., 2022h).

3.6.1.10 WOMEN ARTISAN RELATED STRATEGY

For artisan-related strategies, please refer to Figure 3.2.

3.6.1.11 IMPLEMENTATION STRATEGY

If we want to implement a successful marketing strategy, it will be necessary to create awareness and capacity workshops in various parts of the country. This will help empower small entrepreneurs and female-owned businesses in the manufacturing of jute diversification and promote a green and sustainable approach. It is important to have a specific training program and collaborate with NGOs, SHGs, and local federations for product design (Alawat & Renu, 2018).

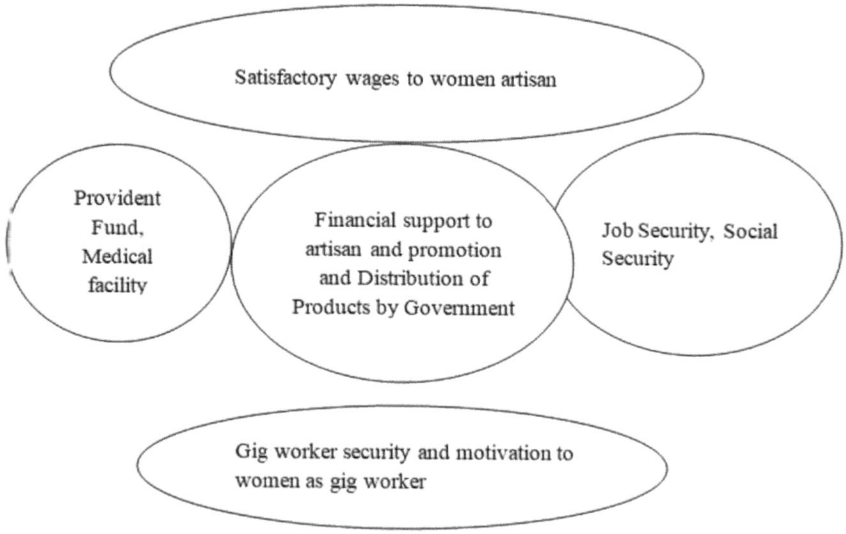

FIGURE 3.2 Women's artisan strategies.
Sources: Figure designed by author.

3.6.1.12 OPENING OF DATA CENTERS AND INSTITUTIONS FOR THE HANDICRAFT INDUSTRY

There is a need for the establishment of Indian institutions dedicated to the management of handicrafts, similar to IIM (Indian Institute of Management) and IIT (Indian Institute of Technology) for design and training in handicrafts. The government should consider the creation of a data center that consolidates information related to artisans, craft design, marketing centers, facilitation centers, textile parks, and design banks. This centralized resource will greatly assist in the study of handicraft development and marketing by artisans and students (Chawla, 2013).

3.7 RESULTS

From the above discussion, it is clear that marketing strategies can play a crucial role in the handicraft industry (ODOP) and the well-being of women involved in this sector, whether as workers or entrepreneurs. These strategies can address the challenges faced by the handicraft industry, increase the income of rural women and urban residents of Uttar Pradesh, whether they are involved in the handicraft sector or not. Additionally, these strategies can help alleviate the migration problem experienced during and after the pandemic crisis, as well as mitigate the diverse impact of COVID-19 that people have faced for the past two years, ultimately boosting the Indian economy.

Uttar Pradesh has great marketing potential and fulfills the criteria of demand, supply, and production of handicrafts in the country, providing employment opportunities for many families involved in the handicraft sector. These strategies will contribute to the promotion of the handicraft sector, artisan involvement, and the empowerment of women in informal and small industries. Based on the findings, it is evident that if these strategies are implemented by women entrepreneurs and supported by reputable NGOs working in this sector, the situation of women entrepreneurs can be significantly improved, and ODOP can create more opportunities for artisans in the handicraft sector. However, it is crucial for the government to provide better facilities, especially for women entrepreneurs.

3.8 CONCLUSION

Although the pandemic has created more problems for people due to the lockdown in all industries, including MSME, the handicraft industry has also been greatly affected. Many people have lost their jobs and returned home empty-handed. The marketing sector has also been disrupted, leaving women artisans and workers unemployed. However, women artisans and marketing managers are striving to improve their situation by marketing and transforming into entrepreneurs. ODOP plays a key role in providing opportunities for women to create jobs, upgrade their skills, and start businesses at the local and global levels. Despite this, handicraft industries and female-based businesses in Uttar Pradesh face significant challenges due to the worldwide corona virus crisis. This sector suffers from a lack of awareness about handicraft education, limited market intelligence, inadequate exposure to new technologies, low capital, and an insufficient institutional framework.

KEYWORDS

- **digital gig workers**
- **handicraft sector**
- **institutional framework**
- **micro-enterprises**
- **one district, one product**
- **strategies**
- **vocal for local**
- **women artisan related strategy**
- **women entrepreneurs**

REFERENCES

Agra, H. R., (2017). Role of government and non-government organizations for production and marketing of Chikankari craft in Lucknow. *Small Enterprises Development Management and Extension, 43*(2), 654–675.

Alawat, V., & Renu, (2018). An analysis of growth and association between labor productivity and wages in Indian textile industry. *Management and Labor Studies, 43*(1, 2), 78–87. doi: 10.1177/0258042x17745182.

Al-Dafri, S., & Alasania, M., (2020). Impact of total quality management, organizational excellence, and entrepreneurial orientation on organizational performance: Empirical evidence from the public sector in UAE. *Benchmarking: An International Journal, 27*(9), 2497–2519.25.

Anand, A., & Bibha, S., (2020). Impact of COVID-19 pandemic on micro, small, and medium enterprises (MSMEs): India report microwave consulting, Lucknow. *Indian Journal of Marketing, 14*(2), 76–78.

Chawla, M., (2013). Customers' (students') perceptions about 7ps of higher education marketing mix. *Asian Journal of Multidisciplinary Studies, 1*(5), ISSN: 2321–8819.

CII, (2020). *CII Sets Fund for MSME to Tackle COVID-19*. https://www.manufacturingto-dayindia.com/people/6997-cii-sets-up-fund-for-MSME-to-tackle-covid-19(accessed on 11 January 2023).

Ghosal, I., & Prasad, B., (2020b). Delineate the transverse environment trade of the handicraft Industry market space: A small empirical survey in Kolkata City, W.B. *ANVESHAK, International Journal of Management, 10*(1). **doi**: 10.15410/aijm/2021/v10i1/157645.

Ghosal, I., Prasad, B., & Behera, M. P., (2020a). Delineating the exchange environment of handicraft Industry from market space to marketplace: An inclusive map for RURBAN development. *Paradigm, 24*(2), 133–148. https://doi.org/10.1177/0971890720959533.

Gitanjali, G., & Nivedita, G., (2021). Impact of COVID-19 on the traditional handicrafts of Assam: A study of *Japi* making craft. *Journal of Rural Development.* http://dx.doi.org/10.25175/jrd%2F2021%2Fv40%2Fi1%2F166502.

GoI, (2020). *COVID-19: Finance Commission's Advisory Pane suggests Support to Small Biz, NBFCs, Press Information Bureau, Government of India.* https://pib.gov.in/PressReleasePage.aspx?PRID=1617900 (accessed on 11 January 2023).

GoI, (2020). *Index of Industrial Production, Monthly, Ministry of Statistics and Program Implementation.* Government of India.

Jafari, S. V., Dutta, D. K., Ferraris, A., & De Giudice, M., (2020). Internationalization business and marketing processes in under-supported policy evidence from Italian SMEs. *Business Process Management Journal, 26*(5), 1055–1074.

Jochhon, K. G., & Indrajit, G., (2019). A research paper to identify the potential cause of the decline in handloom saree and also to opportunities for revival. A case study of handloom tent saree of Begampur hugely districts. *ANVESHAK, International Journal of Management, 8*(1). doi: 10.15410/aijm/2019/v8i1/.

John, R., (2017). Business trends in selected areas of Indian banking: An overview. *IOSR Journal of Business and Management,19*(4), 46–50. doi: 10.9790/487x-1904034650.

Kapoor, (2018). *Registration of Micro, Small, and Medium Enterprises (MSMEs) in India (Udyog Aadhaar Memorandum).*Office of Development Commissioner, Ministry of MSME, Government of India.

Khan, W. A., & Amir, Z., (2013). Study of handicraft marketing strategies of artisans in Uttar Pradesh and its implications. *Research Journal of Management Sciences, 2*(2).

Khurana, S., Haleem, A., Luthra, S., Huisingh, D., & Mannan, B., (2021). Now is the time to press the reset button: Helping India's companies to become more resilient and effective in overcoming the impacts of COVID-19, climate changes, and other crises. *Journal of Cleaner Production, 280*(2), 124466.

Magiya, Razvela, Tavana, M., Kowsari, F., & Rezapour, R., (2011). Supply chain resilience: A benchmarking model for vulnerability and capability assessment in the automotive industry. *Benchmarking: An International Journal, 27*(6), 1929–1949.

Nassir, H. S., (2015). Indian handicrafts: An overview of growing or depleting? *IOSR Journal of Business and Management (IOSR-JBM).*

Sahoo, P., (2020). COVID-19 and Indian economy: Impact on growth, manufacturing, trade and MSME sector. *Global Business Review, 21*(5),1159–1183.https://doi.org/10.1177/0972150920945687.

Shankar, Y. U., Ashish, T. M., Tripathi, S., & Tripathi, R., (2022a). Impact of lockdown during COVID-19 pandemic on micro small and medium enterprises (with special reference to Indian handicraft sector): A study on determinants and exit strategies. *Academy of Marketing Studies Journal, 27*(S1), 1–13.

Tripathi, M. A., Tripathi, R., & Yadav, U. S., (2022). The gig economy: Workers, work, and platform perspective. *Journal of Positive School Psychology, 26*(2), 3434–3450.

Yadav, U. S. Y., Ravindra, T., Mano, A. T., Rajesh, K. S., Gyan, P. Y., & Aliza, (2022b). Entrepreneurial development of artisan in ODOP in Uttar Pradesh to boost economy: Strategies and new approaches towards global handicraft index for socio-economic welfare of artisans. *Asian Journal of Management, Entrepreneurship, and Social Science, 2*(1), 1–17. Retrieved from: http://www.ajmesc.com/index.php/ajmesc/article/view/46 (accessed on 11 January 2023).

Yadav, U. S., & Tripathi, R., (2022f). India's toy industries and markets competition with global toys: An overview of the toy Industry and how the sector is gearing up for an Aatmanirbhar Bharat. *Bank and Policy Journal, 2*(2), 86–110.

Yadav, U. S., Ravindra, T., & Mano, A. T., (2022i). One district one product (ODOP) of Uttar Pradesh: New initiative for developing global handicraft index. *Int. J. Adv. Multidiscip. Res., 9*(2), 1–23.doi: http://dx.doi.org/10.22192/ijamr.2022.09.02.

Yadav, U. S., Tripathi, R., & Tripathi, M. A., (2020). Strategies for development of handicraft sector (small industries) in India. *SEDME (Small Enterprises Development, Management & Extension Journal), 47*(3), 175–193. https://doi.org/10.1177/09708464211037466.

Yadav, U. S., Tripathi, R., & Tripathi, M. A., (2022a). Digital analysis of the transformation of institutions in the knowledge and innovation system of the handmade carpet industry. *SEDME (Small Enterprises Development, Management & Extension Journal), 49*(1), 107–124. https://doi.org/10.1177/09708464221096903.

Yadav, U. S., Tripathi, R., & Tripathi, M. A., (2022e). Exclusive and digital analysis of the transformation of institutions in the knowledge and innovation system of the handmade world carpet industry. *Journal of Positive School Psychology, 6*(3),1135–1160.

Yadav, U. S., Tripathi, R., Tripathi, M. A., & Gyan, P. Y., (2022g). Indian small industries (terracotta of Gorakhpur and Bankura) and women artisan in the digital and COVID-19 era: A case study on the traditional handicraft in Uttar. *International Journal of Economy and Innovation, 5*(2), 359–370.

Yadav, U. S., Tripathi, R., Tripathi, M. A., & Yadav, G. P., (2021). Business strategies for developing Indian handicraft sector (MSME) during post-pandemic COVID 19: Role of artisans as an entrepreneur in boosting economy *Turkish Journal of Physiotherapy and Rehabilitation, 32*(3). ISSN 2651-4451; e-ISSN 2651-446X. www.turkjphysiotherrehabil.org (accessed on 11 January 2023).

Yadav, U. S., Tripathi, R., Tripathi, M. A., Rawat, R., & Kushwaha, J., (2022c). Performance of women artisans as entrepreneurs in ODOP in Uttar Pradesh to boost economy: Strategies and away towards global handicraft index for small business. *Academy of Marketing Studies Journal, 26*(1), 1–19.

Yadav, U. S., Tripathi, R., Yadav, G. P., & Tripathi, M. A., (2022d). Proposal of a global handicraft index for sustainable development: A visionary approach for small Industry and developing strategies for handicraft (rural industry). *European Journal of Sustainable Development Research, 6*(2), em0185. https://doi.org/10.21601/ejosdr/11909.

Yadav, U. S., Tripathi, R., Yadav, G. P., & Tripathi, M. A., (2022h). Proposal of a global handicraft index for sustainable development: A visionary approach for small Industry and developing strategies for handicraft (rural industry). *European Journal of Sustainable Development Research, 6*(2). doi: 10.21601/ejosdr/11909.

CHAPTER 4

Managing (Mis-/) Disinformation Through Big Data Analysis: Resorting to Sustainable Strategies to Counter Infodemic

MANPREET ARORA,[1] ROSHAN LAL SHARMA,[2] VAISHALI DHIMAN,[1] and SUNAINA RATHORE[1]

[1]*HPKVBS, School of Commerce and Management Studies, Central University of Himachal Pradesh, Dharamshala, Himachal Pradesh, India*

[2]*Department of English, Central University of Himachal Pradesh, Dharamshala, Himachal Pradesh, India*

ABSTRACT

The coronavirus pandemic has ravaged and caused extensive damage to every corner of the globe inhabited by the human population. There is not a single country that has remained unaffected due to the COVID-19 pandemic. Amidst this unprecedented crisis, the sharing of relevant, pertinent, and genuine information about the pandemic, its devastating effects, and strategies to overcome and control it becomes extremely crucial. The mis-/disinformation (false/misleading information with the intention of influencing people) that circulates worldwide through various regulatory bodies and social media is not always reliable. This peculiar blend of pandemic and information glut has been causing immeasurable harm rather than benefiting the suffering masses. The term 'infodemic'

Sustainable Strategic Business Infrastructure Development and Contemporary Digital Practices in Industry 5.0. Indrajit Ghosal, Bikram Prasad, Nilanjan Ray, & Lalit Kumar Joshi (Eds.)

came into existence in 2003 and has assumed considerable relevance, particularly during the coronavirus pandemic. It also implies the rapid flow of information (also known as information glut) through various print and social media channels regarding important issues.

Interestingly, the concept of infodemic intersects with the issue of mis-/disinformation, especially in the era of post-truth. The notion of post-truth represents a departure from truth and verifiable facts, instead relying on popular sentiments and emotions. One of the tactics employed in post-truth politics is the phenomenon of fake news, where truth is compromised at every turn, and falsehood, propaganda, and misperception are preferred. It is worth noting that big data has played a significant role in detecting and addressing disinformation by analyzing it critically and objectively. This proposed chapter aims to analyze misinformation/disinformation and various aspects of the infodemic through the lens of big data analysis. The goal is to develop sustainable strategies to combat the spread of misleading information intentionally shared to misguide, confuse, or distort facts. Existing literature suggests that careful and strategic analysis of big data can effectively counter and mitigate mis-/disinformation, as well as deceptive information disseminated through social, print, and other media channels.

4.1 RESEARCH QUESTIONS

The proposed chapter is qualitative in nature and is based on existing literature available on websites, reports, public, and social media platforms. It also aims to qualitatively find answers to the following questions:

- What is infodemic and how is it related to mis-/disinformation?
- How can we manage mis-/disinformation through big data analysis?
- How can big data help deal with infodemic in its wider perspective?
- What can be the sustainable strategies to counter infodemic?

4.2 RESEARCH METHODOLOGY

As mentioned earlier, this chapter is conceptual and qualitative in nature. It relies on data from secondary sources such as reputable websites and published reports to support the theoretical perspective and opinions. Extensive literature review has been conducted to establish the theoretical aspects of the paper. Additionally, a content analysis of tweets related to 'Infodemic' and 'Misinformation' was performed using the qualitative software NVivo.x64 version (QSR international, product version 20.7.0.1533, accessed on October 11, 2022). Twitter analysis was conducted using the keyword 'misinformation,' resulting in the extraction of 17,623 tweets through the NCapture extension.

4.3 INFODEMIC AND MISINFORMATION

It is widely known and believed that, in the age of information technology, "information is power" just like knowledge. However, this does not necessarily mean that more information equates to more power. On the contrary, it often takes the form of an "infodemic" filled with misinformation, disinformation, and rumors. The World Health Organization (WHO) defines an infodemic as "an excessive amount of information, including false or misleading information, circulating in both digital and physical environments during a disease outbreak" (www.who.int/health-topics/infodemic). An infodemic represents a situation characterized by the uncontrolled spread of information or an overwhelming abundance of all types of information, including fake, low-credibility, and unverified sources. Within this abundance, some information may be accurate, but a significant portion consists of disinformation that not only misinforms but also misleads users by exerting various forms of influence. Infodemic, which mixes and disseminates rumors and fears, is propagated among people through both physical and digital information systems (Tangcharoensathien et al., 2020). This makes it extremely challenging for individuals to rely on and trust information sources when they need them. In an infodemic situation, malicious actors often take advantage, leading to confusion and chaos. Infodemic involves tampering with information to manipulate genuine facts, thereby exposing us to misinformation (Swire-Thompson & Lazer, 2019).

One of the biggest examples of an infodemic and misinformation has been the COVID-19 pandemic. When the virus was detected, there was a flood of information in the form of an infodemic, as the flow of information transmission was enormous and highly elevated. As a result, panic was created over social media, and the infodemic in the form of misleading information overwhelmed the mass media in a massive way (Alam &

Chu, 2020). Similar trends of infodemic and misinformation were seen during other epidemics such as Zika, Ebola, and the yellow fever outbreak (Shahi et al., 2021). To mitigate the COVID-19 infodemic, the WHO initiated their own platform named 'Myth Busters,' which invalidated the misinformation by encouraging tech firms to fight against the spread of false news on their social media platforms (Thomas, 2020).

There were certain physical as well as psychological consequences of misinformation during COVID-19 (Su et al., 2021; Nelson et al., 2020; Moscadelli et al., 2020; Sasidharan et al., 2020; Gupta & Sahoo, 2020). A high amount of confusion was triggered based on unknown fears. As there was no established drug or vaccine for the virus, many researchers in various countries were conducting a variety of experiments, leading to the promotion of inconsistent opinions. A high level of inconsistency and a lot of fear were posted on social media by people around the world, further exacerbating the infodemic (Salari et al., 2020; El-Far Cardo, 2021; Moran, 2020). For example, the usage of Hydroxychloroquine was highly propagated as an effective drug in COVID-19, which was highlighted as misinformation by various medical institutions around the world (Recuero et al., 2020; Englund et al., 2020; Mackey et al., 2021; Haupt et al., 2021; Grimes, 2021; Banerjee & Meena, 2021). At one point in time, even President Donald Trump openly advocated for this drug. To counter misinformation during the pandemic, the WHO and many other organizations around the world had to run various awareness campaigns to combat the risk of false or incorrect information regarding the coronavirus pandemic. These campaigns aimed to address the panic caused by the infodemic and disinformation about COVID-19, and to debunk myths about the spread, diagnosis, and treatment of the virus (Bajinka et al., 2020; Kebede et al., 2020; Shaikh et al., 2020; Naeem et al., 2021; Malik et al., 2022).

Along with the infodemic, there has been an exponential growth of misinformation, which has caused tremendous confusion and paved the way for misleading and incorrect recommendations regarding the prevention of the pandemic. This misinformation has proven to be harmful and has fueled the outbreak even further. In such situations, information professionals play a vital role in disseminating filtered, reliable, and truthful information so that the masses can rectify the misinformation (Aleixandre-Benavent et al., 2020) and deal with it accordingly.

At times, even information from reliable sources has turned out to be misinformation, as seen in the context of COVID-19. In fact, the information disseminated by government agencies worldwide has proven to be inaccurate and has caused more chaos and confusion, resulting in the loss of many lives. Misinformation is not only responsible for spreading false rumors, but it can also accentuate fear, cause socio-cultural conflict, and prompt direct harm (Pennycook & Rand, 2020). Additionally, misinformation may be influenced by local factors. Misinformation is not shared or propagated with the intent to cause harm; it is simply false or misleading information (Datta et al., 2020), which automatically causes severe damage and irreparable harm interpersonally, psychologically, socio-culturally, and economically.

In the case of an infodemic, disinformation escalates rapidly and is further propagated through various interconnected channels such as websites, social media, radio, chat apps, and TV. It is also consumed through other communication channels (Tangcharoensathien et al., 2020). The spread of misinformation generally occurs when there is a scarcity of authentic, true, and reliable information. In certain critical circumstances, when information becomes crucial and unavailable, individuals and communities experience nervousness and anxiety. In most cases, anxiety is a major source of the spread of misinformation. To alleviate the stress caused by this, individuals gather unreliable information from various sources and start sharing it based on assumptions that may or may not be true. This, in turn, leads to the spread of fake news, rumors, and mis-/disinformation. Due to the lack of fact-check mechanisms, a vigilant legal system, and stringent law enforcement agencies, infodemic/misinformation circulates unhindered, causing unimaginable harm.

This infodemic cannot be completely eliminated, but it can be managed with the help of information experts and professionals. However, managing the infodemic has become an extremely challenging task, particularly after the rise of social media, which has accelerated the spread of disinformation. The rapid flow of information through social media makes it daunting to control, as consumers tend to consume misinformation without taking the time to filter or stop its spread over the internet or social media platforms.

The term 'infodemic' was coined by the WHO to replace 'epidemic' due to the prevalence of disinformation during the global COVID-19 pandemic. The infodemic during the pandemic has been primarily in the form of rumors and conspiracy theories originating from certain geopolitical regions. A study has highlighted the development of various "metonymic conspiracy theories" to compensate for hardships and alleviate public fears during this unprecedented crisis. It also suggested that COVID-19 was a bioweapon that escaped from a foreign lab (Stephens, 2020). Social media platforms play a significant role in circulating misinformation from low-credibility sources. Yang et al. (2021) found that Twitter and Facebook are the two major platforms responsible for spreading low-credibility information. They also discovered that verified and high-profile accounts are the most influential spreaders of misinformation. Furthermore, they observed that the widespread dissemination of low-credibility content on both platforms is coordinated.

4.4 BIG DATA AND INFODEMIC

Big data is a large volume of data that contains structured, unstructured, and semi-structured data. When we think of big data, we tend to refer to the high usage of technology facilities for storing, analyzing, and managing bulk data. There can be a variety of big data sources from which voluminous data can be extracted, including transaction processing systems, customer databases, documents, emails, mobile apps, and social networks. Big data is characterized by five features: volume, value, variety, velocity, and veracity. The size and complexity of big data are so large that traditional data management tools cannot handle it efficiently. Social media is one of the biggest generators of big data, with over 500 terabytes of data generated every day through photo and video uploads, message exchanges, and comments. This data poses a serious challenge as it is mostly unstructured and heterogeneous, containing combinations of text files, images, and videos. The volume of big data is its peculiar feature, related to the enormous size of the data. Furthermore, data can come in a variety of forms such as emails, photos, videos, PDF files, etc. The term 'velocity' refers to the speed at which data is generated from various sources like mobile devices, social media platforms, networks, and business processes.

One of the biggest challenges in handling big data is managing its variability, which often arises from the data and can hinder effective data handling and management processes. Integrating big data technologies with business processes can offer various advantages to businesses. That is why the importance of data analytics is increasing day by day. Through big data, information managers not only measure certain parameters but also gain more insights into their businesses, enabling them to make informed and better decisions. Big data related to a particular business is typically stored in computer databases and analyzed using software programs specifically designed to handle large and complex datasets. Some examples of big data applications include forecasting demand, understanding consumer purchase intentions, identifying at-risk patients in healthcare systems, predicting crime, and optimizing inventory, among others.

For the management of the "infodemics" created with the help of big data, Chiolero (2022b) recommends a "slow data" culture with slow information production to facilitate public health surveillance. Managing such "big data" generated from different populations and different mediums at a fast pace is highly problematic. Bechmann (2020) points out that to tackle disinformation, media infrastructure must be efficiently built up to ensure stricter flow of information in a decentralized network. Additionally, it has been suggested that data access policies should be consistent to analyze harmful manipulated content (Yang et al., 2021).

Any misinformation that is circulated with the help of IT can impede disaster management. Shen et al. (2022), after analyzing the data related to COVID-19 available on social media through text mining, have concluded that moderate information can increase public compliance with policies during disastrous times. However, a large volume of information being shared through IT can create "information fatigue" and may confuse the public in complying with the policies made for disaster management. In checking the flow of misinformation, "fact-checking" professionals help in countering misinformation, but the fact-checks provided by them get limited visibility to the concerned audience on social media platforms (Micallef et al., 2020). So, the concerned audience or the crowd users of social media must help in sharing information about these fact-checking mechanisms and fact-check platforms.

Mackey et al. (2020) analyzed the big data available on social media related to pharmaceutical products and selling claims for COVID-19 treatment. They found that cybercrime posed the biggest challenge for regulatory authorities in applying "data intelligence" to prevent the dissemination of harmful content on social media.

There have been certain studies that have investigated both the positive and negative impacts of big data generated through social media during the coronavirus pandemic. Social media has been identified as a major source of misinformation and fear-stimulation during the pandemic. However, it can also have a potentially positive impact on public health (Chew & Gunasekeran, 2021). Thus, to reduce the spread of the pandemic, online resources have been found to be the most productive and helpful for the government to identify public attention towards public health messages, so that the best policy reactions can be undertaken (Tran et al., 2021). Furthermore, a study highlighted the potential of "data dashboards" to end the pandemic by continuously reporting data related to public health surveillance generated by health data scientists to policymakers (Chiolero, 2022a). Artificial intelligence (AI) and big data are thus enormously effective for the management of the pandemic due to their potential to convey real-time information about emergencies along with the requisite health interventions (Bragazzi et al., 2020).

4.5 TWITTER ANALYSIS

The various themes identified by NVivo software can be broadly categorized into two main categories: 'misinformation' and 'spreading misinformation.' The twitter dataset revealed a total of 17,623 tweets related to misinformation, with the keyword 'misinformation' being mentioned 6,245 times. Within the category of 'misinformation,' the most frequently mentioned themes were vaccine misinformation (245 mentions), COVID misinformation (149 mentions), misinformation policy (182 mentions), spreading misinformation (80 mentions), blatant misinformation (52 mentions), and combating misinformation (46 mentions), among others.

In the major theme of 'spreading misinformation,' the keyword 'spreading misinformation' appeared in 2,740 mentions within the dataset. Other keywords associated with this theme included dangerous spreader (58 mentions), spreading life (34 mentions), spreading vaccine misinformation (11 mentions), spreading hate (10 mentions), and spreading disinformation (7 mentions).

By using the tweets extracted by NVivo software, a sentiment analysis was performed to assess public sentiments towards "misinformation." The content posted on the social media platform "Twitter" can help us understand whether public views are positive or negative towards the keyword "misinformation." The sentiment analysis revealed that out of the mentions in the tweets related to misinformation, 3,293 tweets were very negative, while 5,610 tweets were moderately negative. Only 1,754 tweets were moderately positive, whereas 710 were very positive. The following hierarchy Table 4.1 displays the four quadrants of sentiment analysis, indicating that the majority of tweets on misinformation are negative or neutral.

TABLE 4.1 Sentiment Analysis

	A: Very negative	B: Moderately negative	C: Moderately positive	D: Very positive
1: Files\\misinformation-Twitter Search~Twitter	3293	5610	1754	710

Source: Software Output, NVivo.

4.6 WORD CLOUD ANALYSIS

The above word cloud shows that the most frequently used words in the tweets extracted from Twitter are "misinformation," "spreading," and "misleading." The tweets captured with the help of "N capture" are evaluated here through a word cloud, which shows different ways to express views about "misinformation" with related words such as 'misleading,' 'lies,' 'terrifies,' and 'wrong.' It can also be further evaluated that social media has been a great platform to spread misinformation. Media in any form is a great source of information to the public, but it may also be criticized severely for spreading misinformation (Figure 4.1).

4.7 GEOGRAPHICAL ANALYSIS

The geographical map above illustrates the distribution of tweets related to the keyword "misinformation" across different continents. The highest number of tweets, 612 in total, were found in South America, specifically from the USA. Following that, the second highest number of tweets, 276 in total, were observed in Colorado, North America. Europe, particularly the UK, ranked third with 233 tweets on "misinformation," followed by Australia (60), Asmara in Africa (45), India in Asia (39), and the remaining tweets originating from other countries around the world. Interestingly, six tweets were even posted from Antarctica. The majority of tweets on "misinformation" were contributed by the public in South America, North America, Europe, and Australia (Figure 4.2).

FIGURE 4.1 Word cloud extracted through NVivo.

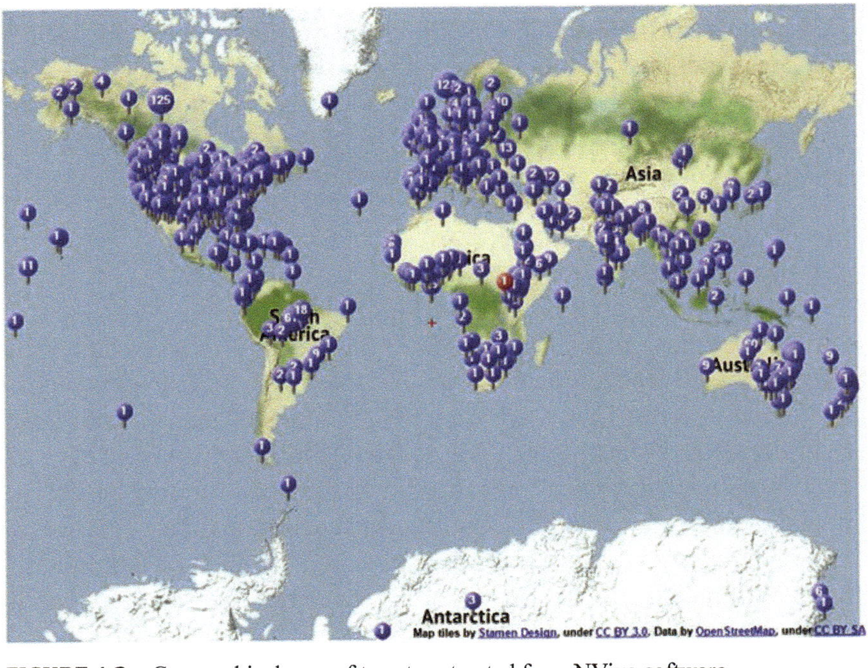

FIGURE 4.2 Geographical map of tweets extracted from NVivo software.

4.8 STRATEGIES TO COUNTER INFODEMIC

The multifaceted nature of misinformation in every field presents significant challenges, making it difficult to counter. It is nearly impossible to expect a single intervention to provide a solution to this problem. To develop effective strategies for dealing with misinformation, it is crucial to first establish a comprehensive understanding. These strategies may involve community-based interventions, collaborative learning, increased social support, web-based interventions, or legal interventions. The development of a robust system to address misinformation will depend on the nature and severity of the problem. Additionally, web-based solutions and social media will play a crucial role in countering infodemics.

1. **Prevention Strategies:** Most of the infodemics are based on manipulating the psychology of the general public. They are designed to create a sense of trust in misleading information. This has the power to shape public discourse and make it extremely difficult to counter. The key to addressing this issue lies in the institutional structures and organizational power. If these structures and power relationships are not committed to addressing the problem, it can spiral out of control. Prevention starts from within the organization, by regularly sensitizing stakeholders and making them aware of their responsibility in preventing negative consequences. Any identified loopholes should be addressed as a priority, as even the slightest oversight can lead to a serious breach of information, resulting in misinformation and ultimately an infodemic. Therefore, preventing misinformation starts with creating a healthy organizational climate that emphasizes information security, effective handling of information, and regularly educating stakeholders on the importance and sensitivity of treating information as a valuable and sensitive asset.
2. **Use of Big Data and Analytics:** Those organizations that are heavily reliant on data, big data solutions, and artificially intelligent systems have found them to be indispensable for efficient data handling and management. Digital platforms serve as the primary source for collecting data, which can be utilized to effectively combat misinformation and fake news. The vast and voluminous data can be processed effectively using big data solutions,

artificially intelligent solutions, and machine learning interventions to prevent mishandling of information and address existing issues, such as during the COVID-19 pandemic. Data extracted by big data solutions from digital platforms played a crucial role in the spread of infodemics and disease control. The inherent ambiguity of infodemics necessitates the use of big data-driven models and advanced analytics technology for crisis management. The detection of cases and forecasting of the pandemic were significant contributions of big data analytics during the pandemic, where the use of heat maps and movement data to track population flow played a vital role in disease control.

3. **Awareness and Sensitization of the Stakeholders:** There have been very grave implications found in the sharing of fake news, such as political instability, public panic, conflict development, and the spread of hatred. Therefore, there is a great need to counter the development of misinformation through various awareness strategies. Social media stakeholders should be adequately informed about the sharing of misinformation through advertisements, workshops, seminars, conferences, and other forms of sensitization. As social media is considered a strong producer of fake news, infodemic, and misinformation, it is also seen as an influential platform on the internet that can act as a powerful weapon for raising awareness and mitigating misinformation.
4. **Legal Strategies:** Various international organizations, such as the WHO, believe that if infodemics are related to diseases or health misinformation, they should be strongly addressed through legal measures. These legal provisions should be supported by government-led awareness campaigns, with a focus on enhancing digital and health literacy among the public. Literacy and awareness are powerful tools to manage the increased spread of fear and panic caused by false or misleading information during crises. By combining proper awareness with legal interventions, it is possible to counter the excessive production of misinformation from multiple sources. Social media is recognized as a major platform for the rapid spread of misinformation, which can create panic and mistrust among the public. Therefore, utilizing social media as a strategy can help combat the spread, speed, and severity of misinformation.

4.9 CONCLUSION

Infodemics and big data are the buzzwords of today's era. Along with this, there has been an exponential growth of misinformation that creates unnecessary public panic, conflicts, and false rumors, which ultimately cause harm. This qualitative study revolves around the analysis of big data and misinformation in the form of tweets extracted from Twitter using the "NCapture" function of NVivo. The analysis was conducted using NVivo software on a Twitter dataset of 17,623 tweets. Using a word cloud, the study has identified major themes such as vaccine misinformation, COVID misinformation, blatant misinformation, etc. Additionally, sentiment analysis revealed the most negative tweets (3,293) and the very positive tweets (710). This study also showed the geographical representation of tweets using a geographical map, where the maximum number of tweets were found in the South America continent (612) and the minimum tweets were six from the Antarctica continent. Lastly, strategies like prevention, awareness, legal measures, and public engagement have been discussed to counter the infodemics.

KEYWORDS

- **big data**
- **big data analysis**
- **counter infodemic**
- **geographical analysis**
- **misinformation**
- **Twitter analysis**
- **word cloud analysis**

REFERENCES

Alam, N., & Chu, C., (2020). COVID-19 pandemic: Tackling 'infodemics' through an integrated one health-social science approach. *Proceedings of the Royal Society of Queensland, 128*, 98–111.

Aleixandre-Benavent, R., Castelló-Cogollos, L., & Valderrama-Zurián, J. C., (2020). *Information and Communication During the Early Months of COVID-19: Infodemics, Misinformation, and the Role of Information Professionals, 29*(4). Research in information and communication. doi: https://doi.org/10.3145/epi.2020.jul.08.

Bajinka, O., Jaw, M., & Kitila, F. L., (2020). Busting myths about SARS-CoV-2 viral pandemic to non-medical personnel. *Journal of Oncology Research, 2*(2).

Banerjee, D., & Meena, K. S., (2021). COVID-19 as an "infodemic" in public health: Critical role of the social media. *Frontiers in Public Health, 9*, 610623.

Bechmann, A., (2020). Tackling disinformation and infodemics demands media policy changes. *Digital Journalism, 8*(6), 855–863. https://doi.org/10.1080/21670811.2020.1773887.

Bragazzi, N. L., Dai, H., Damiani, G., Behzadifar, M., Martini, M., & Wu, J., (2020). How big data and artificial intelligence can help better manage the COVID-19 pandemic. *International Journal of Environmental Research and Public Health.*

Chew, A. M. K., & Gunasekeran, D. V., (2021). Social media big data: The good, the bad, and the ugly (Un) truths. *Frontiers in Big Data, 4*, 4–7. https://doi.org/10.3389/fdata.2021.623794.

Chiolero, A., (2022a). Data do not speak by themselves but will help end the pandemic. *The BMJ, 376*, 33594289. https://doi.org/10.1136/bmj.o161.

Chiolero, A., (2022b). How infodemic intoxicates public health surveillance: From a big to a slow data culture. *Journal of Epidemiology and Community Health, 76*(6), 623–625. https://doi.org/10.1136/jech-2021-216584.

Datta, R., Yadav, A. K., Singh, A., Datta, K., & Bansal, A., (2020). The infodemics of COVID-19 amongst healthcare professionals in India. *Medical Journal Armed Forces India, 76*(3), 276–283.

El-Far, C. A., Kraus, T., & Kaifie, A., (2021). Factors that shape people's attitudes towards the COVID-19 pandemic in Germany—The influence of media, politics and personal characteristics. *International Journal of Environmental Research and Public Health, 18*(15), 7772.

Englund, T. R., Kinlaw, A. C., & Sheikh, S. Z., (2020). Rise and fall: Hydroxychloroquine and COVID-19 global trends: Interest, political influence, and potential implications. *ACR Open Rheumatology, 2*(12), 760–766.

Foutz, N. Z., & Li, B., (2022). Impediment of infodemic on disaster policy efficacy: Insights from location big data. *SSRN Electronic Journal*, 1–34. https://doi.org/10.2139/ssrn.4119720.

Grimes, D. R., (2021). Medical disinformation and the unviable nature of COVID-19 conspiracy theories. *PLoS One, 16*(3).

Gupta, S., & Sahoo, S., (2020). Pandemic and mental health of the front-line healthcare workers: A review and implications in the Indian context amidst COVID-19. *General Psychiatry, 33*(5).

Haupt, M. R., Li, J., & Mackey, T. K., (2021). Identifying and characterizing scientific authority-related misinformation discourse about hydroxychloroquine on twitter using unsupervised machine learning. *Big Data & Society, 8*(1), 20539517211013843.

Kebede, Y., Birhanu, Z., Fufa, D., Yitayih, Y., Abafita, J., Belay, A., & Ambelu, A., (2020). Myths, beliefs, and perceptions about COVID-19 in Ethiopia: A need to address information gaps and enable combating efforts. *PloS One, 15*(11).

Mackey, T. K., Li, J., Purushothaman, V., Nali, M., Shah, N., Bardier, C., Cai, M., & Liang, B., (2020). Big data, natural language processing, and deep learning to detect and characterize illicit COVID-19 product sales: Infoveillance study on Twitter and Instagram. *JMIR Public Health and Surveillance*, *6*(3), 1–17. https://doi.org/10.2196/20794.

Mackey, T. K., Purushothaman, V., Haupt, M., Nali, M. C., & Li, J., (2021). Application of unsupervised machine learning to identify and characterize hydroxychloroquine misinformation on Twitter. *The Lancet Digital Health*, *3*(2), e72–e75.

Malik, J. A., Musharraf, S., Safdar, R., & Iqbal, M., (2022). Myths and misconception of COVID-19 among hospital sanitary workers in Pakistan: Efficacy of a training program intervention. *BMC Health Services Research*, *22*(1), 1–8.

Micallef, N., He, B., Kumar, S., Ahamad, M., & Memon, N., (2020). The role of the crowd in countering misinformation: A case study of the COVID-19 infodemic. *Proceedings – 2020 IEEE International Conference on Big Data, Big Data, 2020*, 748–757. https://doi.org/10.1109/BigData50022.2020.9377956.

Moran, P., (2020). Social media: A pandemic of misinformation. *The American Journal of Medicine*, *133*(11), 1247.

Moscadelli, A., Albora, G., Biamonte, M. A., Giorgetti, D., Innocenzio, M., Paoli, S., & Bonaccorsi, G., (2020). Fake news and COVID-19 in Italy: Results of a quantitative observational study. *International Journal of Environmental Research and Public Health*, *17*(16), 5850.

Naeem, S. B., Bhatti, R., & Khan, A., (2021). An exploration of how fake news is taking over social media and putting public health at risk. *Health Information & Libraries Journal*, *38*(2), 143–149.

Nelson, T., Kagan, N., Critchlow, C., Hillard, A., & Hsu, A., (2020). The danger of misinformation in the COVID-19 crisis. *Missouri Medicine*, *117*(6), 510.

Pennycook, G., & Rand, D. G., (2020). Who falls for fake news? The roles of bullshit receptivity, overclaiming, familiarity, and analytic thinking. *Journal of Personality*, *88*(2), 185–200.

Recuero, R., Soares, F., & Zago, G., (2020). *Polarization, Hyperpartisanship and Echo Chambers: How the Disinformation About COVID-19 Circulates on Twitter.* SciELO Preprints.

Salari, N., Hosseinian-Far, A., Jalali, R., Vaisi-Raygani, A., Rasoulpoor, S., Mohammadi, M., & Khaledi-Paveh, B., (2020). Prevalence of stress, anxiety, depression among the general population during the COVID-19 pandemic: A systematic review and meta-analysis. *Globalization and Health*, *16*(1), 1–11.

Sasidharan, S., Singh, D. H., Vijay, S., & Manalikuzhiyil, B., (2020). COVID-19: Pan (info) demic. *Turkish Journal of Anesthesiology and Reanimation*, *48*(6), 438.

Shahi, G. K., Dirkson, A., & Majchrzak, T. A., (2021). An exploratory study of COVID-19 misinformation on Twitter. *Online Social Networks and Media*, *22*, 100104.

Shaikh, K., Memon, A., Memon, I. A., Laghari, Z. A., & Memon, A. M., (2020). Awareness regarding coronavirus pandemic among the population of Sindh, Pakistan: A cross-sectional study. *Sukkur IBA Journal of Computing and Mathematical Sciences*, *4*(1), 28–36.

Stephens, M., (2020). A geospatial infodemic: Mapping Twitter conspiracy theories of COVID-19. *Dialogues in Human Geography*, *10*(2), 276–281. https://doi.org/10.1177/2043820620935683.

Su, Z., McDonnell, D., Wen, J., Kozak, M., Abbas, J., Šegalo, S., & Xiang, Y. T., (2021). Mental health consequences of COVID-19 media coverage: The need for effective crisis communication practices. *Globalization and Health, 17*(1), 1–8.

Susumu, C., Takashi, I., Kenji, I., Asei, I., Hiroto, K., McElwain, K. M., & Masahiro, S., (2021). Social media infodemics and social distancing under the COVID-19 pandemic: Public good provisions under uncertainty. *Global Health Action, 14*, 1. doi: 10.1080/16549716.2021.1995958.

Swire-Thompson, B., & Lazer, D., (2019). Public health and online misinformation: Challenges and recommendations. *Annual Review of Public Health, 41*, 433–451.

Tran, H. T. T., Lu, S. H., Tran, H. T. T., & Van Nguyen, B., (2021). Social media insights during the COVID-19 pandemic: Infodemiology study using big data. *JMIR Medical Informatics, 9*(7), e27116.

Tangcharoensathien, V., Calleja, N., Nguyen, T., Purnat, T., D'Agostino, M., Garcia-Saiso, S., & Briand, S., (2020). Framework for managing the COVID-19 infodemic: Methods and results of an online, crowdsourced WHO technical consultation. *Journal of Medical Internet Research, 22*(6), e19659.

Thomas, Z., (2020). *WHO Says Fake Coronavirus Claims Causing Infodemic*. BBC News.

Trishnika, C., Gireesh, K. S., Rohit, K., Krithika, S., Gadadasu, S., & Yogesh, D., (2021). Impact of the COVID 19 infodemic on knowledge, attitude, and social behavior in India: A mixed method survey. *The International Journal of Intelligence, Security, and Public Affairs, 23*(3), 197–215. doi: 10.1080/23800992.2021.1956777.

Yang, K. C., Pierri, F., Hui, P. M., Axelrod, D., Torres-Lugo, C., Bryden, J., & Menczer, F., (2021). The COVID-19 Infodemic: Twitter versus Facebook. *Big Data and Society, 8*(1). https://doi.org/10.1177/20539517211013861.

Zarocostas, J., (2020). How to fight an infodemic. *The Lancet,395*(10225), 676.

CHAPTER 5

Impact of COVID-19 on Vegetable Sales and Consumption and Its Effects on the Health Sector in Kathmandu for Sustainable Management

INDRAJIT GHOSAL,[1] MEEN BAHADUR KARKI,[2] and MAHIMA KARKI[3]

[1]*Associate Professor, School of Management and Commerce, Brainware University, Barasat, Kolkata, West Bengal, India*

[2]*School of Management and Commerce, Poornima University, Jaipur, Rajasthan, India*

[3]*School of Management, Tribhuvan University (SOM TU), Kathmandu, Nepal*

ABSTRACT

The main aim of this investigation is to determine how the COVID-19 pandemic has affected vegetable sales and consumption in Kathmandu, specifically at Nepal's Kalimati Fruits and Vegetable Market. It also aims to understand the substitutions that residents of Kathmandu valley have made in their kitchens to compensate for the shortage of vegetables. A total of 102 visitors to the central fruit and vegetable market were surveyed using structured questionnaires. The findings revealed that during the peak COVID hours, there was only 20–40% of the usual supply of vegetables available, and the prices of vegetables had increased by 40–75%. Many locals replaced green vegetables with dals, paneer, and dried beans,

Sustainable Strategic Business Infrastructure Development and Contemporary Digital Practices in Industry 5.0. Indrajit Ghosal, Bikram Prasad, Nilanjan Ray, & Lalit Kumar Joshi (Eds.)

along with locally preserved veggies. To ensure sustainable management in times of crisis like the COVID-19 pandemic or natural shortages, the majority of respondents recommended preserving vegetables during the peak production season. The COVID-19 lockdown resulted in the closure of outpatient services in hospitals and the shutdown of many peripheral healthcare institutions, leading to a several-month interruption of routine vital healthcare services. Personal protective equipment (PPE) for medical professionals was also unavailable. The Kathmandu Valley, as well as Nepal as a whole, experienced high rates of maternal and infant mortality, various types of malnutrition, and the burden of mortality and morbidity from both non-communicable and communicable diseases. The effectiveness of COVID-19 vaccination in controlling the spread of the virus was also found to be ineffective.

5.1 INTRODUCTION

Green vegetables and fruits are daily consumable goods that are consumed by end users in their kitchens. The market mostly sells fruits and vegetables such as ginger, pomegranate, mangoes (different varieties), red potato, Indian red potato, Mude red potato, white potato, cauliflower (local), hybrid cucumber, lemon, asparagus, banana, green chili (varieties), dried chili, carrot, turnip, tomato, and Knol-Khol (gongylodes), Bean (local and hybrid), snake gourd, sugar beet, mushroom varieties, lettuce, Indian big tomato, watermelon, muskmelon, fresh fish, bitter gourd, green leafy mustard, parsley, spinach, mint, Red Nasik Onion, green onion, ripe pumpkin, cabbage, broccoli, green soybean, eggplant (brinjal), okra (gumbo, lady's finger), jackfruit, capsicum (sweet chili), radish, Indian papaya, orange, mandarin, apple, red cabbage, green broad leaf mustard, dried and green garlic, lychee, bottle gourd, sweet potato, hybrid radish, celery, coriander leaf, bamboo shoot, and green round pumpkin (Kalimati Market, n.d.).

In Kathmandu, the Kalimati Fruits and Vegetables Market is operated by the Kalimati Fruits and Vegetables Market Development Board (KFVMDB), a government organization. It is the largest fruits and vegetables market in the country, located in Kalimati, Kathmandu. The market was established through a partnership between the United Nations Capital Development Fund (UNCDF) and the then-Kalimati Wholesale Market project. An amount of 4.6 million rupees was allocated for the construction and furnishing of market facilities in Kalimati. In 1955, the project was transformed into an independent entity under the name Kalimati Fruit and Vegetable Wholesale Market Development Board (KFVWMDB), in accordance with the Development Board Act of 1957. The board is also constructing a fruit and flower market on its own 4-acre land in Naubishe and Dhanding. Additionally, it has 11 different outlets in various parts of Kathmandu, including Kalimati, Baphal, Swayambhu, Banasthali, Gongabu, Bashundhara, Tinkune, Gopikrishna, Lainchaur, Ratnapark, and Koteshwor (Kalimati Market, n.d.).

Based on the 2011 Census, Kathmandu City alone has a population of 1,744,240. However, it is believed that the unofficial number might be much higher (over 2,500,000) compared to this census data. Additionally, there is a retail market where small business owners offer their goods. In the market, there are 425 wholesalers, 65 retailers, and 27 fish businesses. The Vegetable Crops Survey of Nepal 2009-2010, conducted by the Central Bureau of Statistics of the Nepali Government, mentioned that in 2070, the total market import was 220,600.268 metric tons, including 80.46% vegetable produce, 12.76% fruit produce, 4.63% masala produce, 1.85% dry fruit produce, and 29% other agro-produce (Kalimati Market, n.d.).

Produce price fixing is often accomplished through negotiation among farmers/suppliers and traders. The Board is not involved in determining the price of produce. Most of the produce is supplied nationally, with 24.37% imported from India and 1.67% from China (Mapping vegetable markets and vendors in Kathmandu Metropolitan City, 2017).

To adhere to social distancing measures, the Kalimati vegetable mart, which supplies fresh food to houses in Kathmandu, has temporarily closed as mandated by the Kathmandu Valley Forum of DAOs (District Administrative Officer) to reduce the spread of COVID-19 in Nepal.

Generally, the wholesale market, semi-wholesale market, retailers' market, and street vendors are the marketing intermediaries in the Nepalese Fruits and Vegetables Market. It is estimated that in Nepal, there are over 50 commonly grown crops and over 247 different types of farmed vegetables. The most prevalent vegetable crops grown in Nepal include cauliflower (33,172 ha), tomato (19,724 ha), cabbage (14,306 ha), pumpkin (9,757 ha), cucumber (8,634 ha), eggplant (8,172 ha), okra (7,473 ha), hot pepper/chili (7,007 ha), and bitter gourd (4,250 ha). In Nepal, the government has built 8 wholesale markets, 4 retail markets, 33 trade centers for fruits and veggies, 21 collecting centers, and so on. Some of these markets are run by the private sector (Awasthi, 2007).

It is estimated that approximately 50 to 60% of the total vegetable demand in Kathmandu is met by vegetable production in hilly regions such as Kavre, Dhading, Bhaktapur, Lalitpur, Makwanpur, Kathmandu, Nuwakot, Sindhupalchok, Gorkha, Dolakha, Rasuwa, etc. The remaining vegetables are sourced from Terai districts, as well as from India and China. Additionally, vegetables are also being cultivated in rooftop kitchen gardens of cement-plastered houses (Mapping vegetable markets and vendors in Kathmandu Metropolitan City, 2017).

COVID-19 has caused loss of life, disrupted livelihoods, disrupted daily life, and dominated global political discussions for over two years (Yves Leterme, former Prime Minister of Belgium, Chair of the Global Commission on Democracy and Emergencies, and member of Club de Madrid (Kantipur Post, 2022, Vol. 03, No. 01)). As the acute phase of the pandemic comes to an end, the researcher also intends to assess what COVID-19 has revealed about the ability of Kathmandu residents to handle similar situations. Binaya Shrestha, Deputy Director of the Kalimati Fruits and Produce Market, stated that the market receives 300–450 tons of vegetables per day. Due to the lockdown, the supply has decreased by 25.33%, and the market has only received 280 tons of vegetables. At Balkhu, the second-largest vegetable market in the valley, daily exports of 200–250 tons of produce have recently declined (Prasain, 2020).

Global human life and health systems are currently facing enormous challenges and hazards as a result of the COVID-19 (coronavirus infectious disease) pandemic. Current infection control and management mainly rely on public health preventive measures, such as lockdown and social isolation, to stop disease transmission in the absence of effective treatments for COVID-19. By the end of March 2020, more than 100 nations had implemented partial or total lockdowns in an effort to halt the spread of the COVID-19 illness, resulting in millions of people being confined to their homes. Motivated by this, this chapter aims to address the following objectives: to determine consumer expenses on vegetables during normal days and during the main COVID-19 affected days. How can the differences in vegetable costs for consumers be determined between normal days and COVID-19 days? What caused the price increase for fruits and vegetables during the COVID-19 period? Which vegetables did consumers substitute during the peak of the COVID-19 outbreak? How has COVID-19 affected health, and how effective has the vaccine been in stopping the virus in the Kathmandu Valley?

5.2 REVIEW OF LITERATURE

Permanent markets are typically open throughout the day. These markets sell all types of fruits and vegetables. They act as wholesale markets for vegetables and fruits, with small vendors purchasing from them for their businesses. Generally, the public purchases food at these permanent markets for larger meals prepared during feasts and festivals (Mapping vegetable markets and vendors in Kathmandu Metropolitan City, 2017).

Levi, Paulson, & Perakis (2020) found that access impacts a household's expenditure on fruits and vegetables, driven by the frequency of purchases. This effect is observed in homes with low nutritional values and those located within a mile of each other.Karki & Ghosal (2022) concluded that the SERVQUAL model, used to shape the survey on consumer satisfaction, indicated that tangible variables and services collectively influence consumer satisfaction.

Ghosal et al. (2021) conducted research and discovered that food consumption has a significant impact on nature, people, and global health. The present study aims to examine the individual and environmental factors that influence customer decisions to purchase green items online. According to Genkin & Mikheev (2020), the report from the Food and Agriculture Organization (FAO), World Trade Organization, and World Health Organization (WHO) highlights the threat of food devastation caused by the ongoing COVID-19 pandemic, with a risk of a global "food shortage" due to disruptions in the trade industry's supply chain.

According to Madhav Timilsina, president of the Consumer Rights Investigation Forum, the large vegetable and fruit market will see inflated prices if only wholesale commerce is permitted. Only a few traders and people have access to the wholesale market as many consumers rely on the retail market (Prasain, Kalimati fruits and vegetables market stops retail sales, 2021).

Adhikari et al. (2021) found in their study on "COVID-19 impacts on agriculture and food systems in Nepal: Implications for SDGs" that the lockdown and commuting constraints have had negative effects, raising concerns about the achievement of SDGs 1 and 2, especially in the existing food-aid-dependent and vulnerable regions.

Leterme (2022): For more than two years, COVID-19 has taken lives, destroyed livelihoods, disrupted daily routines, and dominated political discussions around the world (Yves Leterme, the former Prime Minister of Belgium, the Chair of the Global Commission on Democracy and Emergencies, and a member of Club de Madrid (Kantipur Post, 2022, Vol. 03, No. 01)).

Devendra et al. (2021) conducted research entitled "Impact of COVID-19 on health services utilization in Province-2 of Nepal: a qualitative study among community members and stakeholders." They found that COVID-19 has significantly worsened Nepal's decades-long progress in maternal and pediatric healthcare. Maternal mortality and neonatal deaths have abruptly surged by more than three times during the epidemic.

5.3 MATERIALS AND METHODS

5.3.1 RESEARCH METHODOLOGY

A descriptive research design was planned. Structured questionnaires were used to collect primary data from market gate purchasers of the Kalimati market, health centers, hospitals, and village health units. Secondary data were collected from research papers, brochures, magazines, and daily newspapers.

Semi-structured qualitative interviews were conducted with purposefully selected individuals as part of the investigation into the health implications of COVID-19. Both virtual and in-person interviews were conducted in the participants' native languages. Thematic network analysis was employed to examine the data. Statistical tools such as the Chi-Square test and percentages were utilized to analyze and interpret the data and results.

5.3.2 POPULATION AND SAMPLE

The population consists of all potential buyers of fruits and vegetables from the Kalimati market. A total of 102 respondents were selected to gather information using structured questionnaires, and their comments were recorded. In order to test the effectiveness of the vaccine, approximately 1,000 individuals at various hospital entrances were asked whether they had been vaccinated or infected.

5.3.3 SOURCES OF DATA

This study primarily focuses on collecting primary data through structured questionnaires via online surveys, as well as field visits. Secondary data were obtained from literature reviews, papers, booklets, etc.

5.3.4 DATA ANALYSIS TOOLS

The data was analyzed using Excel and SPSS. Chi-square test and frequency test were carried out for the data analysis part.

5.4 RESULTS AND DISCUSSION

This section presents the analysis of the collected data from the consumer survey. Therefore, this chapter aims to fulfill the objectives set at the beginning of the study.

5.4.1 DEMOGRAPHIC PROFILE

Table 5.1 clearly states that the number of female respondents was 61, accounting for 59.8%, and the number of male respondents was 41, accounting for 40.2%. The age class with the highest number of purchasers was 26–35 years, accounting for 56%, indicating that this group of purchasers used to go to the market for purchasing. This was followed by the age group of 36–45 years, accounting for 24.8%.

TABLE 5.1 Demographic Profile of Respondents

Demographic Variable		**Frequency**	**Percentage (%)**
Gender	Male	41	40.2
	Female	61	59.8
Age	16–25	14	12
	26–35	44	56
	36–45	27	24.8
	Above 45 years	17	7.2
Total		**102**	**100**

5.4.2 EXPENSES (BEFORE AND IN COVID-19 AFFECTED DAYS) IN VEGETABLES PURCHASES (RS.) ON PER DAY BASIS

Table 5.2 showed that 45 respondents used to spend Rs. 101–150 in vegetables at normal days as compared to the expenditure of Rs. 151–200 in vegetables purchasing during COVID hit days consisting of 44.12%. Their expenditure on vegetables has increased by 40%. Around 38 respondents spent Rs. 50–100 on vegetables during normal days, compared to Rs. 101–150 during the COVID-19 hit days, which accounted for 37.25% of the respondents. Their expenditure on vegetables has increased by 66%. Similarly, 11 respondents spent Rs. 151–200, compared to Rs. 201–250 on everyday vegetable expenditure, which accounted for 10.78% of the respondents. There was a 20% increase in the price for this category of respondents. Around 8 respondents spent Rs. 200–250 and more than Rs. 250 on vegetable buying during both normal and COVID hit period. They constituted 7.85% of the respondents, and there was only an 11% hike in vegetable prices (Table 5.2).

TABLE 5.2 Expenses in Vegetable Purchase Before and in COVID-19 Affected Days

Before Affect	Affected Day	Frequency	Percentage (%)	Percentage Increase in Expenses
50–100	101–150	38	37.25	66
101–150	151–200	45	44.12	40
151–200	201–250	11	10.78	20
200–250	Above 250	8	7.85	11
Total		**102**	**100**	

5.4.3 REASONS FOR PRICE INCREASE IN VEGETABLES

The attendants were asked to provide details in Table 5.2 regarding the reasons for the increase in vegetable prices. Out of 102 respondents, 49 clearly specified that the price hike was due to the unavailability of vegetables, which required customers to spend more time and energy to find and purchase them, accounting for 48.03% of the respondents. The second largest group, comprising 42 respondents or 41.18%, believed that vegetables were simply expensive (Table 5.3).

TABLE 5.3 Reasons for Price Increase

Statements	Frequency	Percentage (%)
Vegetables were not available, and customers spent more time and energy to find and purchase vegetable.	49	48.03
Vegetables were expensive.	42	41.18
Supplied vegetables were rotten and inadequate in supply.	6	5.88
Vegetables were available in distant market and in retail sales only.	5	4.91
Total	**102**	**100**

5.4.4 SUBSTITUTION FOR VEGETABLES WHEN VEGETABLES WERE NOT AVAILABLE

Table 5.3 clearly depicts that 56 of the respondents substituted vegetables with dried beans, Rajma, and Paneer, constituting 54.90%. This was followed by chicken and eggs substitutions, each by the same number of consumers' families, which is 23 respondents constituting 22.55% (Table 5.4).

TABLE 5.4 Substitution for Vegetables

Item	Frequency	Percentage (%)
Chicken	23	22.55
Mutton	0	0
Eggs	23	22.55
Dried beans, rajma, and paneer	56	54.90
Total	**102**	**100**

5.4.5 SOLUTION FOR FUTURE CRISIS

In the answer to this query, Table 5.4 clearly shows that 48 of the respondents opined that preservation solves the vegetable shortage in such situations, constituting 47.06% of the total. This was followed by the opinion of maintaining roof kitchens in urban areas, supported by 35 respondents, accounting for 34.32%. Additionally, around 10 respondents (9.80%) emphasized the importance of vegetable storage to address the crisis. Around, 9 respondents (8.82%) highlighted the significance of kitchen gardens in rural areas as a means to address the crisis (Table 5.5).

TABLE 5.5 Vegetable Shortages for Kathmandu-based Customers in the Possible Future Crisis

Statements	Frequency	Percentage (%)
Vegetable preservation	48	47.06
Vegetable storage	10	9.80
Roof kitchen in urban areas	35	34.32
Kitchen garden in rural areas	9	8.82
Total	**102**	**100**

5.4.6 CHI-SQUARE TEST

For this 10 days' supply record as provided by the board and required supply were recorded and Chi-square test was applied (Table 5.6).

TABLE 5.6 Observed and Expected Import of Vegetables from Different Points to the Kalimati Market

Days	O = Observed Supply (Tons)	E = Expected Supply (Tons)	O–E	(O–E)2	((O–E)2)/E
1	523	725	–202	40,804	56.28
2	660	725	–65	4,225	5.83
3	582	725	–143	20,449	28.21
4	584	725	–141	19,881	27.42
5	703	725	–22	484	0.67
6	732	725	7	49	0.067
7	761	725	36	1,296	1.79
8	668	725	–57	3,249	4.48
9	470	725	–255	65,025	89.69
10	294	725	–431	1,85,761	256.22
Total = 10 days	**5,977**	**7,250**	–	–	**470.657**

Source: Kalimati Fruits and Vegetable Market Board dated 2021-07-14.

The results showed that the calculated Chi-square value of 470.657 at 9° of freedom and a 5% level of significance is greater than the tabulated (critical) value.

The Chi-square value of 16.919 at a 5% level of significance and 9° of freedom also supports this interpretation. Therefore, it can be concluded that the supply of vegetables to the Kalimati market significantly decreased during the active contagion period. This justifies consumers' claims of unavailability of vegetables.

Furthermore, the same 102 respondents were asked about the effects of COVID-19 on their health and their families, as well as the measures they suggest for mitigating the effects of such pandemics in the future. Around 68 respondents (67%) reported that almost every member of their family was infected, and three respondents lost a family member – one in each family. They mentioned that hospitals were closed, health centers lacked health technicians, transportation was completely halted, schools and colleges were closed indefinitely, government offices operated with only 20% of the staff on a rotating basis, and grocery outlets and fruit and vegetable markets were shut down. Everyone was confined to their homes. Both COVID-19 patients and those with other illnesses died without receiving treatment.

Around 1,000 people who were visiting hospitals in the Kathmandu Valley were surveyed to determine if they had been vaccinated and if they experienced any adverse effects after vaccination. A chi-square test was conducted (Table 5.7(a)).

TABLE 5.7(a) Chi Square Test for Vaccine Effectiveness

	Vaccinated	**Not Vaccinated**	**Total**
Affected	107 (102)	193 (198)	300
Not Affected	233 (238)	467 (462)	700
Total	**340**	**660**	**1,000**

The numbers within the bracket indicated expected calculated numbers (Table 5.7(b)).

TABLE 5.7(b) Chi-Square Test for Vaccine Effectiveness

Observed No. (O)	**Expected No. (E)**	**(O–E)**	**$(O–E)^2$**	**$\Sigma(O–E)^2/E$**	**χ^2Value**
107	102	5	25	0.2450	0.5304
193	198	–5	25	0.1263	
233	238	–5	25	0.1050	
476	462	5	25	0.0541	
			Total	**0.5304**	

This clearly indicated that vaccine was not found effective in controlling COVID-19.

5.5 CONCLUSION

From the study it was concluded that:

1. Vegetables and fruits prices were significantly hiked.
2. During the COVID period, there was a shortage of vegetables and fruits, and the supply was significantly reduced.
3. People have substituted vegetables and fruits with chicken, preserved vegetables, and pulses due to the shortage in supply.
4. The respondents under study also suggested having kitchen gardens in rural areas and rooftop kitchens in urban and metropolitan areas.
5. The child and maternal mortality rates were found to be higher during COVID due to a shortage of medicines, the closure of health service centers, a lack of health technicians, insufficient Personal Protective Equipment (PPE), and delays in vaccination.

5.6 MANAGERIAL IMPLICATIONS

Studying the results of this research, managers in the concerned fields will be able to address problems similar to situations like COVID-19 in the future. The findings also provide guidelines to government officials and help formulate national-level policies to deal with these types of challenging times. Consumers will also benefit from these research findings as they learn the importance of preserving vegetables when they are abundantly available during the main production season. This study also teaches them to consume only the optimum amount and avoid wasting cooked food, while also encouraging the use of preserved vegetables as substitutes for fresh ones in their kitchens. The study results also emphasize the need for government emergency preparedness and swift action during crises. Additionally, they help bring important issues into focus for the government. Furthermore, the results assist health personnel and institutions in implementing safety measures, preventive and curative remedies, and formulating health strategies and policies to combat situations similar to COVID-19.

5.7 FUTURE SCOPE OF THE STUDY

People from all over Nepal have been settling in Kathmandu since the advent of democracy in 1990. They purchase fruits and vegetables from this market, which is under government control. Other large sellers also buy in large quantities from this market. In the future, researchers can study fruit sales and consumption in this market during crises like COVID-19, as well as in other metropolitan cities like Biratnagar and Pokhara, by taking large sample sizes to enhance the applicability of the research findings. They can use a large sample size to study health-related issues and include more health centers, hospitals, and paramedical fields to further validate the study.

5.8 LIMITATION OF THE STUDY

Convenient sampling has been used with a sample size of 102 respondents. The sample size is small due to the short time frame available for the research. Therefore, generalization cannot be done. For health-related studies, only 1,000 samples were taken into consideration, but a larger sample size would have been preferable.

KEYWORDS

- **consumption**
- **COVID-19**
- **data analysis tools**
- **data sources**
- **green vegetables**
- **health effects**
- **Kathmandu market**
- **sustainability**
- **sustainable management**

REFERENCES

About Kalimati Market. Retrieved from: Kalimati Market. https://kalimatimarket.gov.np/ (accessed on 11 January 2023).

Adhikari, et al., (2021). COVID-19 impacts on agriculture and food systems in Nepal: Implications for SDGs. *Elsevier Public Health Emergency Collection, 1*(1), 1–30.

Awasthi, B., (2007). *Relevance of Market Information System to Environment Protection.*

Devkota, A. R., Dhakal, D. D., Gautam, D. M., & Dutta, J. P., (2014). Assessment of fruit and vegetable losses at major wholesale markets in Nepal. *International Journal of Applied Sciences and Biotechnology (IJASBT), 2*(4), 559–562.

Genkin, A. S., & Mikheev, A. A., (2020). Influence of coronavirus crisis on food Industry economy. *Foods Raw Mater.*

Ghosal, I., Gupta, B., Dey, M., & Prasad, B., (2021). Impact of outlining factors towards online green products: An empirical investigation on consumers' awareness. *International Journal on Recent Trends in Business and Tourism (IJRTBT), 5*(1), 1–7.

Ghosal, I., Prasad, B., & Gupta, B., (2022). Restructuring the green consumerism through e-commerce portals: A Behavioral congruence during post-COVID-19. In: *Future of Work and Business in Covid-19 Era*, 89–99. *Journal of Agriculture and Environment* (Vol. 8, No. 1, pp. 705–726.

Karki, & Ghosal, (2022). Analysis of customer's satisfaction towards service quality of hotel Industry in Nepal. *International Journal on Recent Trends in Business and Tourism, 6*(1), 15–25.

Leterme, Y., (2022). *COVID-19's Lessons for Democracies–The Kantipur Post* (p.14). The National Daily News Paper.

Levi, Paulson, & Perakis, (2020). Fresh fruit and vegetables consumption: The impact of access and value. *MIT Sloan Research Paper, 1*(1), 10–41.

Mapping Vegetable Markets and Vendors in Kathmandu Metropolitan City, (2017). Retrieved from: Resource Center for Primary Health Care (RECPHEC).

Prasad, B., & Ghosal, I., (2021). Forecasting buying intention through artificial neural network: An algorithmic solution on direct-to-consumer brands. *FIIB Business Review*, 23197145211046126.

Prasain, K., (2020). *Vegetables to be Sold at 11 Designated Sites as Kalimati Bazaar is Shuttered.* The Kathmandu Post. Retrieved from: https://kathmandupost.com/29/2020/04/09/vegetables-to-be-sold-at-11-designated-sites-as-kalimati-bazaar-is-shuttered (accessed on 11 January 2023).

Prasain, K., (2021). *Kalimati Fruits and Vegetables Market Stops Retail Sales*. The Kathmandu Post. Retrieved from: https://kathmandupost.com/money/2021/05/12/kalimati-fruits-and-vegetables-market-stops-retail-sales (accessed on 11 January 2023).

Sarkar, M. D., & Ghosal, I., (2018). Impact of social media in small business industry: A study on Facebook. *International Journal on Recent Trends in Business and Tourism (IJRTBT), 2*(3), 33–39.

Singh, D. R., Sunuwar, D. R., et al., (2021). Impact of COVID-19 on health services utilization in province-2 of Nepal: A qualitative study among community members and stakeholders. *BMC Health Services Research, 1*(1), 25–37.

White, J. T., (1999). *Market Infrastructure Planning – A Guide for Decision Makers*. FAO Agricultural Services Bulletin.

Zhu, F., & Zhang, X., (2010). Impact of online consumer reviews on sales: The moderating role of product and consumer characteristics. *Journal of Marketing, 74*(2), 133–148.

CHAPTER 6

Review of Industry 5.0: Evolution and Future Advancements

PANKAJ ANAND,[1] JITENDRA KUMAR,[1] and SUBHADEEP MUKHERJEE[2]

[1]*Department of Management Studies, Sikkim Manipal University, Sikkim, India*

[2]*Ramaiah University of Applied Sciences, Bengaluru*

ABSTRACT

The study discusses the different phases of the Industrial Revolution, from 'Industry 1.0 to Industry 5.0.' The researchers have examined the various advancements achieved during each Industrial Revolution. The study will also assess the benefits and key challenges of Industry 5.0. Additionally, the concept of VUCA (volatile, uncertain, complex, and ambiguous) was thoroughly studied, considering the ever-changing dynamics of today's world. The researchers have also examined the impact of Information Technology and its role in observing Industry 5.0. Furthermore, various visions and perspectives provided by different authors for Industry 5.0 were assessed. The implications of this study for managers and academics were also examined. This chapter explores the evolution of Industry 5.0 and the various components required for its advancements.

6.1 INTRODUCTION

The human race is currently experiencing the fourth Industrial Revolution. Since the 1st Industrial Revolution in the 17th century until today, the industrial sector has undergone significant changes, largely attributed

Sustainable Strategic Business Infrastructure Development and Contemporary Digital Practices in Industry 5.0. Indrajit Ghosal, Bikram Prasad, Nilanjan Ray, & Lalit Kumar Joshi (Eds.)

to advancements in the technological sector. Each era of the Industrial Revolution has brought about advancements that distinguish it from its predecessor. This ever-changing and evolving nature of the market, from Industry 1.0 to Industry 5.0, can be perfectly summarized by the concept of VUCA, which stands for "Volatility, Uncertainty, Complexity, and Ambiguity." The changes and evolution taking place are due to the VUCA concept. The VUCA concept can be most accurately described as quick and unpredictable changes, a lack of proper standards, and a chaotic nature. The term VUCA describes the dynamic changes in the market that directly or indirectly impact the functioning of any firm (Nowacka & Rzemieniak, 2021).

Below listed are few important attributes of each industrial revolution (Figure 6.1):

1. **Industry 1.0:** This era marks the usage of water or steam power for engines for mechanization and mass production (Agolla, 2018).
2. **Industry 2.0:** This era is marked by the usage of electricity and the start of assembly line production for mass production from the early 20th century (Baygin et al., 2016).
3. **Industry 3.0:** This era is marked by the usage of programmable logic controllers (PLCs), also known as programmed electronic devices, in place of electrical and mechanical machines (Baygin et al., 2016). Additionally, the utilization of computers for automation has become more prominent.
4. **Industry 4.0:** This is the era of people, products, and devices all interacting in real time using high production models of automated digital products and services (Zhou et al., 2015). Various cyber-physical systems (CPS) were prevalent during this era.
5. **Industry 5.0:** In this era, all advanced technologies such as robots, artificial intelligence (AI), and augmented reality are expected to be used for the advancement of human life, whether it be in industry, healthcare, or other daily human activities (Martynov et al., 2019). Society has progressed towards mass customization and cyber-physical cognitive systems.

After all the automations in Industry 4.0, people are now seeking customization or personalization, which could be a potential direction for implementing AI in automated systems.

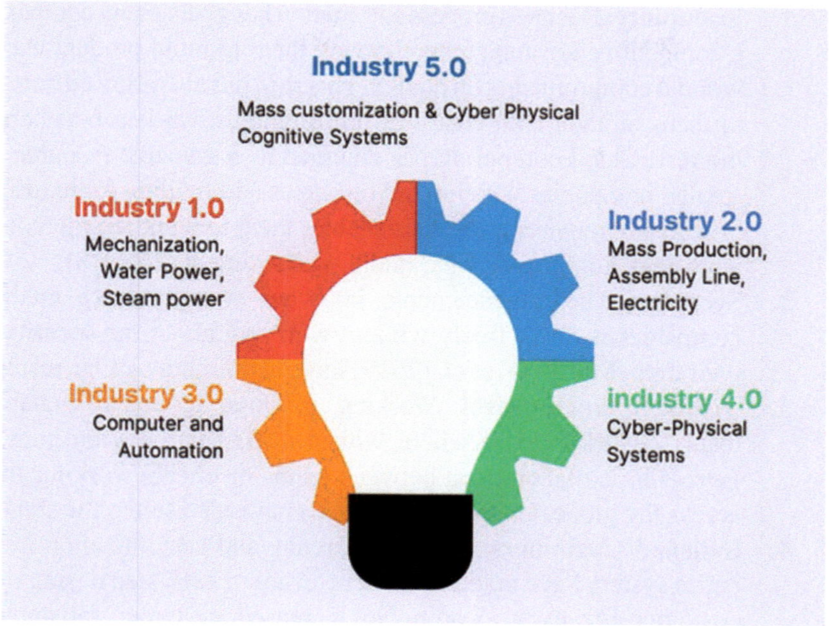

FIGURE 6.1 Industry 5.0: Publicizing the epoch of intelligent automation–Intellias.
Source: Adapted from: Martynova (2019).

Industry 4.0 is characterized by the usage of the internet to facilitate human-to-human as well as human-to-machine communication. The major advantage is uninterrupted, cheaper, and endless communication (Roblek et al., 2016). Rüßmann et al. (2015) proposed nine pillars of Industry 4.0: 'horizontal and vertical system integration, autonomous robots, simulation, big data analytics, cloud computing, industrial internet of things (IIoT), augmented reality, cybersecurity, and additive manufacturing.' According to Martynov et al. (2019), the four important aspects of Industry 4.0 are: 'internet of things (IoT), smart factories, CPS, internet of services (IoS),' along with six more additive technologies such as 'IIoT and CPS, virtual reality, 3D printing, AI, big data, and collaborative robots (CoBot).' The three key features of Industry 4.0, as given by Sharma et al. (2020), are 'innovative and digital business models, integration of horizontal and vertical value chains, and digitized service and product offerings.'

As of now, humans are experiencing Industry 4.0, wherein we witness various advancements. Discussing the benefits of Industry 4.0, Sharma et al. (2020) pointed out the following advantages:

1. **Scalability:** Business process automation has given many companies a competitive advantage by allowing them to build products faster without compromising on quality. This shift has also allowed humans to focus on their own creativity and strengths, while robots handle the rest. Cloud computing has emerged as a valuable resource for smaller companies with limited resources and funding for hardware or software requirements. This enables them to work remotely from anywhere, aligning with the findings of Ardito et al. (2018).
2. **Security:** The advancements in cyber security have enabled companies to work freely without worrying about the security of their data. Martynov et al. (2019) have also found similar results.
3. **Visibility and Control:** Working on cloud servers has enabled teams to work from anywhere, with more transparency and control, increasing collaborations between teams or clients working from across the globe. Cabrilo et al. (2020) have also stated the same.
4. **Satisfied Customers:** The transparency and customization in the entire system have made customers aware of each step during their customer life cycle, leading to increased customer satisfaction and overall ease of the process. Ardito et al. (2018) found results consistent with the above.
5. **Customization:** Various advancements, such as commercial 3-D printers and laser cutting machines, have made prototyping and production much more customizable and faster. Everything can now be tailored to the specific requirements and needs of the customers, which aligns with the findings of Roblek et al. (2016).

Industry 3.0 primarily focused on mass production, while Industry 4.0 shifted its focus towards customization. Industry 5.0, on the other hand, is all about complete personalization. Companies should prioritize transitioning to Industry 5.0 to gain a competitive advantage while keeping costs as low as possible (Martynov et al., 2019). With rapid technological advancements, people now prefer customized products, which require collaboration between humans and technology. This is the essence of Industry 5.0. To enhance product customization and improve efficiency, humans and machines will work together (Sharma et al., 2020). The use of CoBots will increase, and Industry 5.0 will also emphasize renewable energy, bioeconomy (the utilization of biological resources to achieve a balance between economy, industry, and ecology), and waste reduction.

6.2 LITERATURE REVIEW

Humans have already experienced three major industrial revolutions and are currently in the midst of the fourth one, known as Industry 4.0. Furthermore, we are rapidly approaching the fifth industrial revolution, referred to as Industry 5.0. According to Henderman et al. (2020), the first industrial revolution (Industry 1.0) involved the transition from hand production to mechanical loom production, utilizing water and steam engines. The second industrial revolution (Industry 2.0) introduced electricity and assembly line production, enabling mass production. Subsequently, the third industrial revolution (Industry 3.0) witnessed the emergence of programmable logic controllers (PLCs), as well as advancements in electronics, IT systems, and automated production lines. Currently, we are in the midst of the fourth industrial revolution (Industry 4.0), characterized by the utilization of cyber-physical systems (CPS), internet of things (IoT), robotics, artificial intelligence (AI), big data, and cloud computing. Looking ahead, we are rapidly approaching the fifth industrial revolution, where human-robot collaboration will become commonplace and the bioeconomy will play a crucial role (Demir et al., 2019). Industry 5.0 will be characterized by the integration and utilization of robotics and AI, with a strong focus on sustainability, renewable resources, and bionics (Figure 6.2).

The German government first mentioned Industry 4.0 as a national strategy in November 2011 and implemented the "High-Tech Strategy 2020 for Germany" (Kagermann et al., 2013). Subsequently, several other countries also launched different initiatives to facilitate the implementation of Industry 4.0. The United Kingdom initiated the "UK CATAPULT –High Value Manufacturing" mission, which involved universities and corporations promoting the use of innovative digital technologies. Since then, several more countries have formalized their own initiatives for Industry 4.0, such as the French "Industrie du Futur," the American "Manufacturing USA," and the Dutch "Smart Industry." These missions provide fiscal benefits, financing, and tax credits to companies that align with the vision of Industry 4.0. The Italian Government launched "The Italian Plan for Industry 4.0" to increase public and private funding for automating and digitizing various processes in businesses and industries (Ardito et al., 2018).

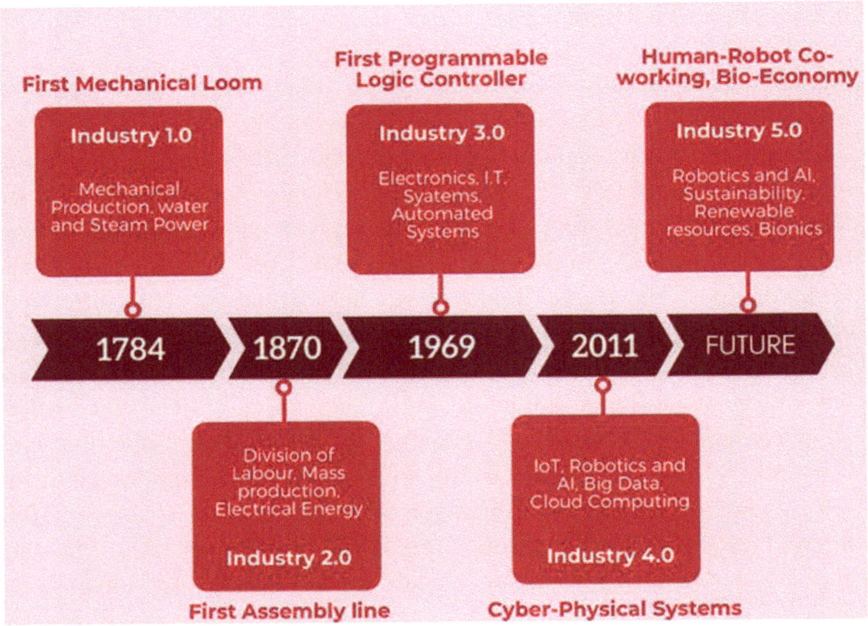

FIGURE 6.2 Evolution of industrial era.

Source: Adapted from: Aslam et al. (2020).

The main concepts of Industry 4.0 are advancements in various Information and Communication technologies, advancements, and automation of industrial technologies, and finally CPSs. The usage of the latest advanced technology is going to be the major differentiator between Industry 4.0 and Industry 5.0. There have been remarkable advancements in inter and intra-organizational collaboration, communication, and delivering innovation at a much cheaper and faster rate, accelerating and facilitating inter-organizational innovations (Sarker et al., 2012; Majchrzak & Malhotra, 2013; Anderson & Hardwick, 2017). Few enablers of Industry 4.0, as given by Menendez et al. (2020), are: Horizontal and Vertical System integration, additive manufacturing, autonomous robots, simulation, industrial internet of things (IIoT), cybersecurity, the cloud, augmented reality, and big data and analytics. Improvement and updating of processes in industries and companies generate a huge amount of data which, in turn, improves the speed of decision making and gives more flexibility in terms of governance (Table 6.1).

TABLE 6.1 Literature Review

SL. No.	Title	Author and Year	Inference	Relevance with Present Work
1.	Industry 5.0 and Smart Cities: A Futuristic Approach	Sharma, I., Garg, I., & Kiran, D. (2020)	The researchers have briefly discussed about Industry 5.0, also giving a background on few topics related to Industry 4.0 such as the technological developments, drawbacks of Industry 4.0 and the underlining structure of smart cities.	Technological advancements and modifications in Industry 5.0 have been studied in detail.
2.	Competencies for Industry 4.0	Hernandez-de-Menendez, M., Morales-Menendez, R., Escobar, C. A., & McGovern, M. (2020)	The researchers have reviewed the various competencies required, the technologies requirements for Industry 4.0.	Industry 4.0 competencies are studied in detail.
3.	Human Capital Mapping for Industry 4.0: Gap and Index	Hendarman, A. F., Sari, F. A., Reza, L. V. I., Damar, M. R., Handayani, M., Zefanya, S., Adinata, S. J. T. & Parera, S. B. (2020)	The research aims to design a framework to map the human capital readiness in order to implement Industry 4.0.	Readiness of human capital for Industry 4.0 has been studied in detail.
4.	Information Technology as the Basis for Transformation into a Digital Society and Industry 5.0	Martynov, V. V., Shavaleeva, D. N. & Zaytseva, A. A. (2019)	The present study aims at analyzing various advanced technologies required for the digitization of Industries so as to ensure a smooth transition to Industry 5.0.	Usage of information technology for transformation to Industry 5.0.
5.	Evaluation of Key Challenges to Industry 4.0 in Indian Context: A DEMATELApproach	Aggarwal, A., Gupta, S., & Ojha, M. K. (2019)	The researchers have attempted to evaluate the various challenges for the realization of Industry 4.0 in the Indian context.	Industry 4.0 has been studied in Indian context.

TABLE 6.1 *(Continued)*

SL. No.	Title	Author and Year	Inference	Relevance with Present Work
6.	Industry 5.0 and Human-Robot Co-working	Demir, K. A., Döven, G., & Sezen, B. (2019)	The researchers have discussed Industry 5.0 in detail various issues related to human-robot co-working and employees' perspective of Industry 5.0.	Usage of technology in Industry 5.0 has been discussed.
7.	Barriers to the Adoption of Industry 4.0 Technologies in the Manufacturing Sector: An Inter-Country Comparative Perspective	Raj, A., Dwivedi, G., Sharma, A., Jabbour, A. B. L., & Rajak, S. (2019)	The researchers have examined various barriers to the implementation of various technologies of Industry 4.0 in manufacturing sector in the context of both developing as well as developed economies.	Barriers to Industry 4.0 are studied in detail.
8.	Towards Industry 4.0 Mapping Digital Technologies for Supply Chain Management-Marketing Integration	Ardito, L., Petruzzelli, A. M., Panniello, U., & Garavelli, A. C. (2018)	In this chapter the researchers have presented a comprehensive view of the innovative efforts undertaken in order to develop the digital technologies and the role these advancements play in improving the supply chain management and marketing.	Several enabling technologies pertaining to Industry 4.0 is studied in detail.
9.	A Maturity Model for Assessing Industry 4.0 Readiness and Maturity of Manufacturing Enterprises	Schumacher, A., Erol, S., & Sihn, W. (2016)	Maturity model for Industry 4.0 is discussed. Overall, 9 dimensions for assessing the maturity of Industry 4.0 has been defined.	Maturity model for Industry 4.0 has been studied.
10.	Industry 4.0: The Future of Productivity and Growth in Manufacturing Industries	Rüßmann, M., Lorenz, M., Gerbert, P., Waldner, M., Justus, J., Engel, P., & Harnisch, M. (2015)	The authors proposed the 9 pillars of Technological advancements, Also, the authors have also quantified the advancements taking Germany as an example.	Various impacts of Industry 4.0 are studied based on manufacturing sector.

Based on the aforementioned studies, researchers have discovered that previous researchers have examined the technological advancements and requirements for Industry 5.0. Additionally, they have thoroughly investigated the readiness level of both humans and industries for Industry 5.0. The researchers have also explored numerous benefits and issues associated with Industry 5.0, and have debated the reasons why it should be embraced as a positive change for the current economy (Table 6.2).

TABLE 6.2 Major Research Gaps Identified

SL. No.	Author Name and Year	Title	Research Gaps Identified
1.	Hendarman, A. F., Sari, F. A., Reza, L. V. I., Damar, M. R., Handayani, M., Zefanya, S., Adinata, S. J. T., & Parera, S. B. (2020)	Human Capital Mapping for Industry 4.0: Gap and Index	This study is a quantitative study only, also, the study was limited to one firm in one sector (manufacturing sector) of Indonesia only.
2.	Schumacher, A., Erol, S., & Sihna, W. (2016)	A Maturity Model for Assessing Industry 4.0 Readiness and Maturity of Manufacturing Enterprises	The study focuses only on the companies working in manufacturing sector, manufacturing the goods physically and in-house using their own machinery, also they have their own already defined customer groups.
3.	Ardito, L., Petruzzelli, A. M., Panniello, U., & Garavelli, A. C. (2018)	Towards Industry 4.0 Mapping Digital Technologies for Supply Chain Management-Marketing Integration	The researchers have used the patent data in order to study innovation dynamics which has its own limitations such as, patent data cannot be accounted for few innovations as, either they are not patentable or it might also be that patenting may not be the best protection tool for such innovations.

From the above-identified research gaps, it is observed that the previous studies are either sector-specific or country-specific, and are primarily theoretical-based studies. Further investigation is needed to explore the implementation of Industry 5.0 in detail, and the study should aim to provide generalizable insights for all sectors and countries.

Few researchers have supported two visions for Industry 5.0: "Human-robot coworking" (Rubmann et al., 2015; Roblek et al., 2016; Martynov et al., 2019; Menendez et al., 2020) and "Bioeconomy" (Demir et al., 2019; Sharma et al., 2020). Human-robot coworking involves the collaboration of humans and robots to complete tasks. Humans focus on design and creativity, while robots handle the actual work (Demir et al., 2019). Moving on to the next vision, Bioeconomy refers to the use of renewable and biologically available resources in modern businesses, aiming to achieve a balance between business, environment, and economy. According to the European Commission, Bioeconomy involves the production of renewable biological resources and the conversion of these resources and waste streams into value-added products, such as food, feed, bio-products, and bioenergy. It encompasses sectors like agriculture, forestry, fishing, food production, pulp and paper, as well as parts of the chemical, biotech, and energy industries. These sectors have significant innovation potential due to their utilization of various scientific disciplines (life sciences, agronomy, environment, food sciences, and social sciences), industrial and enabling technologies (biotechnology, nanotechnology, information technology), as well as information and communication technology (ICT and engineering) and local and tacit knowledge.

Robots, as defined by the Oxford dictionary, are "machines capable of automatically performing a complex series of actions, especially those that can be programmed by a computer." Another definition, provided by the Merriam Webster dictionary, describes a robot as "a humanoid machine that performs various complex actions (such as walking or talking) for a human." These definitions are inherently vague, suggesting that a satisfactory definition has yet to be determined. According to Demir et al. (2019), in order to facilitate human-robot collaboration, proper rules and regulations must be established. Firstly, a clear and legally binding definition of robots is required. While we have a scientific definition of a robot, it is necessary to establish a legal definition that can be applied to businesses. There are several types of automated machines, and a robot is one of them. It is important to have a clear law that regulates automated machines and robots, as well as the businesses that use them. Clear guidelines should be established regarding the types of robots that can be used in different places and industries. The roles and responsibilities of robots should also be clearly defined. In the event of a robotic malfunction, the roles and responsibilities of humans should be specified in advance. Additionally, necessary certifications and authorities should be involved in the process. According to Aggarwal et al. (2019), the government should establish a proper and detailed guideline that provides a clear understanding of all the aforementioned aspects in order to facilitate human-robot co-working, which is essential for the realization of Industry 5.0.

6.3 METHODOLOGY

The study examines the evolution from Industrial Revolution 1.0 to Industry 4.0, as well as future advancements in Industry 5.0, referencing various previous and relevant studies conducted between 2012 and 2020. Further classification based on journals is provided in subsections.

6.3.1 JOURNAL SEARCH STRATEGY

The researchers used the following search strategy to search for relevant papers on Industry 5.0. Various open-source database search engines such as Web of Science, Google Scholar, Publish or Perish, etc., were utilized. The keyword used was Industry 5.0. Based on the results, various journals were searched for relevant papers on Industry 5.0.

6.3.1.1 ACCEPTED JOURNALS

Journals with relevant papers related to Industry 5.0 were selected, and the relevant papers were considered for the study. The list of selected journals for the study is provided in Table 6.3.

TABLE 6.3 Classification Based on Journal Indexing

SL. No.	Journal Name	Indexing
1.	The Journal of Strategic Information Systems	SCIE | SSCI
2.	International Entrepreneurship and Management Journal	Scopus, Q1
3.	International Journal of Production Economics	Scopus, Q1
4.	Business Process Management Journal	Scopus, Q1
5.	Journal of Intellectual Capital	Scopus, Q1
6.	International Journal on Interactive Design and Manufacturing	Scopus, Q2
7.	China Information	Scopus, Q2
8.	International Journal of Manufacturing Research	Scopus, Q4

6.3.1.2 REJECTED JOURNALS

Journals with no papers related to Industry 5.0 were rejected and therefore not considered for the current study. Table 6.4 provides a list of the few journals that were not included in the current study.

TABLE 6.4 List of Journals Not Selected for the Study

SL. No.	Journal Name
1.	International Journal Advance Science Technology
2.	The Service Industries Journal
3.	Academy of Management Review
4.	Economic and Political Weekly
5.	Econometrica: Journal of the Econometric Society

6.3.2 KEY CHALLENGES TO INDUSTRY 5.0

The following key challenges, identified from various literature reviews conducted by the researcher, aim to enhance the rapid adoption of Industry 5.0 in the Indian context.

1. **Legal Issues:** Stricter laws related to data privacy and data security need to be enforced. Additionally, regulations need to be maintained to ensure the safety and security of skilled and unskilled workers (Demir et al., 2019).
2. **Leadership:** Industry 5.0 will bring about a significant shift, and any change must be supported by the leadership or management. Therefore, leadership will play a crucial role in the transition to Industry 5.0 (Aggarwal et al., 2019).
3. **Higher Cost:** Industry 5.0 will require advanced machinery and equipment, which in turn will require training of personnel and consultancy services, resulting in an additional cost for the company. This will be a significant challenge for Indian companies, as the majority of them are small and medium-scale industries (Raj et al., 2019).
4. **Resistance from Workers:** Workers or employees are the main pillars of a company, and as Industry 5.0 will be technologically heavy, unskilled workers have a high risk of losing their jobs.

Therefore, workers or employees may resist the changes required for Industry 5.0 (Aggarwal et al., 2019; Raj et al., 2019).

5. **Security Concerns:** Industry 5.0 introduces the concept of IIoT (industrial internet of things), which involves connecting all the devices within an industry. However, this also brings the risk of data breaches and cybercrimes. Any malfunction can result in significant expenses, and any data breach can have severe consequences for the company (Raj et al., 2019).
6. **Governmental Support:** Innovation has been at the top of the Indian government's agenda for a long time. However, more supportive policies are needed for the proper implementation of Industry 5.0. Supportive policies are required for the wider implementation of new technologies such as IIOT, robotics, adaptive systems, artificial intelligence, etc. The government must also support cross-border trades to facilitate the sharing of new emerging technologies (Demir et al., 2019; Raj et al., 2019).

The future directions and prospects of Industry 5.0, as given by Martynov et al. (2019), are as follows:

i. **New Types of Distributed Computers:** These systems use Artificial Intelligence to self-organize the available resources.
ii. **Internet of Things and People:** IoT (also including industrial internet of things (IIoT)), is the root of automation in both Industry 4.0 and Industry 5.0. It basically connects software to hardware required for automation of the complete system.
iii. **Multi-Agent Systems and Technologies:** Various classical mathematical methods are used to solve complex problems for the advancement of technology and to bring about changes to existing ones.
iv. **Theory of Complex Adaptive Systems:** A system that changes or learns according to the inputs is called an adaptive system.
v. **Ontology and Knowledge Bases:** The knowledge presented on the basis of conceptualization is called Ontology.
vi. **Emergent Intelligence:** The dynamic nature of the system and its unpredictability in the decision-making process are key elements of an emergent intelligent system.
vii. **Evergetics:** It is a part of various management processes prevalent in our society. It also helps in bringing out the "good actions" in the management processes.

viii. **Enterprise Architecture:** In order to achieve success in Industry 4.0 and Industry 5.0, companies need to have a well-designed enterprise architecture. This will also enable them to compete effectively in the digital industry.

Previous researchers (Rubmann et al., 2015; Roblek et al., 2016; Martynov et al., 2019; Menendez et al., 2020) have acknowledged the fact that Human-Robot co-working will play an important role in Industry 5.0. As this will be another unique case, there will be several issues related to human-robot co-working. Bagdasarov et al. (2018) conducted a study on the implications of working with robots. The study considered three sets: Individual, organizational, and robotic agent. The researchers further examined each set in detail and suggested factors that need to be considered for each set.

a. **Factors for Individuals:** Various factors related to an individual's demographic details, along with their knowledge and background, are studied.
b. **Organizational Factors:** The organization's working, work environment, focus on human capital, etc., constitute the organizational factors.
c. **Robotic Agent Factors:** The demographics of the robot, along with its safety and security, constitute the factors of the robotic agent.

We acknowledge the ever-changing dynamics of the world we live in, which gives rise to the concept of VUCA. VUCA combines four different types of challenges into one word: volatility, uncertainty, complexity, and ambiguity. The term VUCA stands for:

1. **Volatility:** Volatility occurs whenever there is a sudden change, and these changes cannot be predicted beforehand. Several factors determine volatility in corporations, including digitization of resources, increased global competition, and continuous innovation in the business framework of the organization.
2. **Uncertainty:** The lack of control over the near future is a significant factor. Various unstable internal and external conditions influence the outcomes, thereby diminishing the certainty of achieving desired results.
3. **Complexity:** It can be seen as a by-product of volatility and uncertainty. Today's corporations are surrounded by chaos and turmoil, which adds to the complexity.

4. **Ambiguity:** Technological advancements have given us a number of choices, with several possibilities. And this is the classic case of ambiguity (Nowacka & Rzemieniak, 2021; Wang et al., 2022).

Nowacka & Rzemieniak (2021) presented the following competencies as an answer to each individual word of the VUCA term, which is presented in the form of Table 6.5. Individuals who are familiar with these terms will be able to have a broader perspective, a better understanding of situations, and react more effectively. They will also be able to communicate in a simpler, clearer, and more transparent manner.

TABLE 6.5 Competencies of the Term "VUCA"

VUCA	Description Competences
Vision	The perfect answer to volatility is having a vision for the project. Vision gives a certain trend and the way out, to volatility.
Understanding	The answer to Uncertainty is the answer to any uncertain situation.
	It is to understand the situation in a better way so as to ensure a simple flow of information.
Clarity	Clarity is the answer to complexity. Clarity is the creation of processes and procedures, which are easier to understand by the employees. The result is, all the employees are aware of what to do in a particular situation, regardless of the chaos in the external environment.
Agility	Agility is the answer to ambiguity. It supports managers in ambiguous conditions. Experimentation allows for accurate identification of correct solutions in each situation. Rather than wasting time in discussing things, it's always better to create a prototype and be being able to test it.

The study found that the VUCA model becomes very important to realize the vision of Industry 5.0. The advancements in technologies and workflows bring along volatility, uncertainty, complexity, and ambiguity. As a matter of fact, change is not always welcomed by all, so these advancements are bound to create ambiguity and complexities in the minds of the workforce. And this is where leadership needs to have a proper understanding and clarity over the technology's advancements, in order to guide the workers. Thus, it becomes imperative for leadership to provide proper training and keep the workforce abreast of newer and advanced technologies. And this is where the answers to the VUCA concept will play a major role.

6.3.3 MANAGERIAL IMPLICATION

Managers are the first level of contact for the people actually working on the ground. The newer and more advanced machines and workflows may face some resistance from the workers' community, and it becomes the duty of the managers to implement the changes in a smoother manner. The vision of Human-Robot Co-working can only be realized with the help of these managers, who will be the first point to face the constraints of Human-Robot Co-working. Thus, the 4 answers to the VUCA concept come into play: vision, understanding, clarity, and agility. These 4 will be the main support systems for managers to implement changes and advances in technologies and align them with the available workforce.

6.3.4 ACADEMIC IMPLICATION

Human Resources is one of the most important pillars in achieving the vision of Industry 5.0. The constantly changing aspects of the world pose a significant threat to achieving the goals of Industry 5.0. Therefore, it is imperative for students to stay updated with the latest market trends, and this is where academic institutions play a crucial role. In the ever-evolving world, educators should stay updated with the recent trends and pass on their knowledge to students, preparing them to face new technologies and workflows when they enter the corporate world. This will enable them to meet the requirements of Industry 5.0.

6.4 CONCLUSION

This study has proposed two visions for Industry 5.0, namely "Human Robot Co-working" and "Bio-Economy." Regardless of whether Industry 5.0 is specifically about human-robot co-working, the concept of human-robot co-working is expected to bring significant changes to organizations. The ultimate goal of Industry 5.0 is to promote the adoption of technological advancements, automate processes, and facilitate collaboration between humans and robots using Information Technology as its core. Additionally, the digitization and automation of firm processes, both vertically (across the entire value chain) and horizontally (across functional areas), will be the main areas of focus. Industries will also prioritize sustainable and environmentally friendly approaches to conducting business.

Also, it has become increasingly important for top-level management in corporations to be aware of the VUCA concept. This concept provides a better way to navigate through upcoming complexities and uncertainties. Proper understanding of the VUCA concept will give a clear picture of upcoming problems and, at the same time, provide better solutions to navigate through them.

The researchers studied various problems regarding human-robot co-working. The researchers discovered that several changes are required in organizations to facilitate human-robot co-working. These problems necessitate changes and innovation in organizational structure, human behavior, work structure, and work environment. It is crucial to establish well-defined rules and regulations to address privacy and cybersecurity concerns. Trust is essential for effective co-working with robots. The definition and categorization of robots should be specified to ensure their proper usage and to define the roles and responsibilities of both robots and humans in case of failure or malfunctioning. Training should be provided to enable effective co-working with robots, and greater awareness needs to be created to foster widespread acceptance of robots.

Additionally, the researchers have studied the various key challenges for the realization of Industry 5.0. They have also provided future directions and prospects for Industry 5.0. Furthermore, the researchers discussed the various implications of Human-Robot Co-working.

The study is limited by its reliance on a small number of previous studies on Industry 5.0. Additionally, the researchers have only considered studies conducted between 2012 and 2020. Future researchers should incorporate more recent studies and utilize data on the implementation of various advancements to assess their success or failure.

KEYWORDS

- **evolution**
- **industrial revolution**
- **Industry 4.0**
- **Industry 5.0**
- **managerial implication**

REFERENCES

Aggarwal, A., Gupta, S., & Ojha, M. K., (2019). Evaluation of key challenges to Industry 4.0 in Indian context: A DEMATEL approach. In: *Advances in Industrial and Production Engineering* (pp. 387–396).

Agolla, J. E., (2018). Human capital in the smart manufacturing and Industry 4.0 revolution. *Digital Transformation in Smart Manufacturing*, 41–58.

Anderson, A. R., & Hardwick, J., (2017). Collaborating for innovation: The socialized management of knowledge. *International Entrepreneurship and Management Journal, 13*(4), 1181–1197.

Ardito, L., Petruzzelli, A. M., Panniello, U., & Garavelli, A. C., (2018). Towards Industry 4.0: Mapping digital technologies for supply chain management-marketing integration. *Business Process Management Journal, 25*(2), 323–346.

Aslam, F., Wang, A., Mingze, L., & Khaliq, Ur. R. C., (2020). Innovation in the era of IoT and Industry 5.0: Absolute innovation management (AIM) framework. *China Information, 11*(2), 2–24.

Bagdasrov, Z., Martin, A. A., & Buckley, M. R., (2020). Working with robots. *Organizational Dynamics, 49*(2), 1–8.

Baygin, M., Yetis, H., Karakose, M., & Akin, E., (2016). An effect analysis of Industry 4.0 to higher education. In: *2016 15th International Conference on Information Technology Based Higher Education and Training (ITHET)* (pp. 1–4).

Cabrilo, S., Dahms, S., Mutuc, E. B., & Marlin, J., (2020). The role of IT practices in facilitating relational and trust capital for superior innovation performance: The case of Taiwanese companies. *Journal of Intellectual Capital, 21*(5), 753–779.

Demir, K. A., Döven, G., & Sezen, B., (2019). Industry 5.0 and human-robot co-working. *Procedia Computer Science, 158*, 688–695.

Hendarman, A. F., Sari, F. A., Reza, L. V. E., Damar, M. R., Handayani, M., Zefanya, S., & Parera, S. B., (2020). Human capital mapping for Industry 4. 0: Gap and INDEX. *International Journal Advance Science Technology, 29*(2), 112–119.

Hernandez-de-Menendez, M., Morales-Menendez, R., Escobar, C. A., & McGovern, M., (2020). Competencies for Industry 4.0. *International Journal on Interactive Design and Manufacturing (IJIDeM), 14*(4), 1511–1524.

Langeveld, J. W. A., (2015). *Results of the JRC-SCAR Bioeconomy Survey*. Biomass Research: Wageningen, The Netherlands.

Majchrzak, A., & Malhotra, A., (2013). Towards an information systems perspective and research agenda on crowdsourcing for innovation. *The Journal of Strategic Information Systems, 22*(4), 257–268.

Martynov, V. V., Shavaleeva, D. N., & Zaytseva, A. A., (2019). Information technology as the basis for transformation into a digital society and Industry 5.0. In: *2019 International Conference "Quality Management, Transport and Information Security, Information Technologies" (IT&QM&IS)* (pp. 539–543).

Martynova, O., (2019). *Industry 5.0: Announcing the Era of Intelligent Automation – Intellias. https://intellias.com/industry-5-0-announcing-the-era-of-intelligent-automation/*

Nowacka, A., & Rzemieniak, M., (2021). The impact of the VUCA environment on the digital competences of managers in the power industry. *Energies,* 185–202.

Raj, A., Dwivedi, G., Sharma, A., De Sousa, J. A. B. L., & Rajak, S., (2020). Barriers to the adoption of Industry 4.0 technologies in the manufacturing sector: An inter-country comparative perspective. *International Journal of Production Economics*, *224*, 1–49.

Roblek, V., Meško, M., & Krapež, A., (2016). A complex view of Industry 4.0. *Sage Open*, *6*(2), 1–11.

Rüßmann, M., Lorenz, M., Gerbert, P., Waldner, M., Justus, J., Engel, P., & Harnisch, M., (2015). *Industry 4.0: The Future of Productivity and Growth in Manufacturing Industries* (Vol. 9, No. 1, pp. 54–89). Boston Consulting Group.

Sarker, S., Sarker, S., Sahaym, A., & Bjørn-Andersen, N., (2012). Exploring value cocreation in relationships between an ERP vendor and its partners: A revelatory case study. *MIS Quarterly*, 317–338.

Schumacher, A., Erol, S., & Sihn, W., (2016). A maturity model for assessing Industry 4.0 readiness and maturity of manufacturing enterprises. *Procedia CIRP*, *52*, 161–166.

Sharma, I., Garg, I., & Kiran, D., (2020). Industry 5.0 and smart cities: A futuristic approach. *European Journal of Molecular & Clinical Medicine*, *7*(8), 2750–2756.

Wang, C., Dai, M., Fang, Y., & Liu, C., (2022). Ideas and methods of lean and agile startup in the VUCA era. *International Entrepreneurship and Management Journal*, 1–18.

Zhou, K., Liu, T., & Liang, L., (2016). From cyber-physical systems to Industry 4.0: Make future manufacturing become possible. *International Journal of Manufacturing Research*, *11*(2), 167–188.

CHAPTER 7

Exploring Cloud Computing Platforms and the Metrics Used to Compare Cloud Service Providers

PARUL DUBEY[1] and ARVIND KUMAR TIWARI[2]

[1]*Department of Information Technology, Shri Shankaracharya Institute of Professional Management and Technology, Raipur, Chhattisgarh, India*

[2]*Dr. C. V. Raman University, Chhattisgarh, India*

ABSTRACT

Because data is growing at an exponential rate in today's society, there is an increasing need for cloud computing services. Cloud computing offers several advantages to developers and service providers. The cloud is responsible for the platform's availability, security, processing, storage, and other features. In this study, the focus is on the cloud computing platform. Everything about cloud computing, including the different types of clouds and the various services they provide, can be found in one convenient location. It also provides a comprehensive comparison of different cloud service providers (CSPs). Both the drawbacks and merits of each strategy are presented in detail to ensure a clear understanding of the whole picture. Various aspects such as the type of service provider, the type of service, and the pricing are taken into account during the comparison process.

Sustainable Strategic Business Infrastructure Development and Contemporary Digital Practices in Industry 5.0. Indrajit Ghosal, Bikram Prasad, Nilanjan Ray, & Lalit Kumar Joshi (Eds.)

7.1 INTRODUCTION

Natural clouds are formed by the accumulation of water molecules, and the term "cloud" in cloud computing refers to networks in a similar way to natural clouds (Arockiam, Monikandan, & Parthasarathy, 2011). Users have the ability to access the various modes of operation available in cloud computing at any given moment. Customers who frequently use cloud computing prefer to utilize a third-party mediator provider instead of building their own physical infrastructure. In cloud computing, computer power, database storage, application software, and other IT resources are made available on demand. In this context, "on-demand" is a crucial term. Additionally, users will be able to pay as they go using a cloud service platform. In other words, they will only be charged for what they have requested when they request it, and they will not be charged again after they have finished using it. As a result, we are able to provide clients with the precise computing resources they desire.

The cloud enables users to customize their service according to their specific requirements. This provides immediate access to all resources without the need to pre-order or provide advance notice. Running applications on a local system is not resource-intensive as the cloud networks handle the majority of the service demand. Consequently, there is a reduced need for hardware and software. To utilize cloud computing, all that is required is a web browser.

7.2 RELATED WORK

Research examines a number of different features of present cloud service delivery systems, as well as the prospective uses for software development that they may have (Parkhomenko, Zalyubovskiy, & Parkhomenko, 2021). The development of new SAP apps, upgrades, or add-ons, as well as the setup of a hybrid IT environment on the SAP Cloud Platform, make it an enterprise cloud platform. The SAP, which consists of the DevOps technique and a different architecture, is evaluated in terms of its capabilities and limitations. The Cloud Foundry environment used in the business application development process for the SAP Cloud Platform is detailed in the studied paper.

It is necessary for organizations to choose the finest cloud service provider (CSP) at the most competitive pricing feasible. According to a study (Trueman, Narayanasamy, & Ashok, 2021), a graph-based strategy for assessing CSPs has been developed and is currently being tested. The initial step involves calculating the correlation between the response times of CSPs. To eliminate spurious connections, a graphical lasso regularization network is constructed with a penalty on each node. The service providers are then ranked based on their degree centrality in order to determine their ranking. Finally, a normalized discounted cumulative gain approach is employed to assess the quality of CSPs.

Considerations for selecting a CSP include security, service pricing, and manageability in multi-cloud environments. A multi-cloud system necessitates a solution to the issue of selecting a CSP. The possible CSP combinations are ranked using a multi-criteria decision-making (MCDM) method for each online application (Ramamurthy, Saurabh, Gharote, & Lodha, 2020). Ranking the CSP combinations based on the solution method, which considers the selection of virtual machines (VMs) and cloud data centers, takes this into account. It is probable that it will not be feasible to choose the ideal CSP combinations for different products, both from a management and cost perspective. Consequently, as a starting point, they proposed an optimization model for choosing CSPs for multiple applications inside a corporation, while taking into consideration budget and ranking. Numerical data experiments are utilized to demonstrate their approaches and come up with fresh perspectives on the subject.

Recent research (Jaithunbi, Sabena, & SaiRamesh, 2021) examines the trustworthiness of a service provider by complementing the standard Genetic Algorithm with Intelligent Rules, which is a novel approach. When analyzing trust, it is recommended to include criteria such as reputation, accreditation, service availability, auditing, and self-assessment, among others. With the aid of these ingenious rules, it is possible to affect the SGA based on a single parameter. In this case, the rules are utilized to forecast unpredictability by using their dynamic solution-providing approach based on the Intelligent Rules, which aids in the provision of solutions to complex dynamic systems, where the bulk of unpredictability occurrences take place. Intelligence Rules and SGA collaborate to assess the trustworthiness of cloud servers or cloudlet instances. Due to their dependability, customers are able to make use of the services without restriction.

The following list summarizes the most significant contributions made by an article in the course of its review (Li, 2021). Since a CSP's revenue, expenses, and profits are all quantifiable based on well-established analytical models, a quantitative method was employed to evaluate the quality and price of their services. It is argued that client satisfaction can be divided into two categories: satisfaction with the price of the service and satisfaction with the quality of the service provided. A closed-form expression of customer satisfaction can be derived from a CSP. In highly competitive cloud computing industries, a non-cooperative game paradigm is required. CSPs interact through the market stability mechanism, which is based on the tactics of other CSPs, and the Nash equilibrium occurs as a result of this interaction. The researcher is working on developing an algorithm for identifying the Nash equilibrium.

A framework for CSPs has been created that automates cloud management services such as traffic monitoring, traffic forecasting, virtual service instance placement, and traffic load balancing, among other functions. Researchers have employed various techniques to monitor and predict traffic in the cloud, including the auto-regressive integrated moving average (ARIMA) model for traffic predictions, the Randomized Weighted Majority Algorithm for virtual service instance placement, and a threshold-based strategy for load balancing (Dharmapriya et al., 2019). The article extensively demonstrates the performance of their cloud management platform, with a particular focus on evaluating the algorithms utilized in the aforementioned modules. To achieve efficient and reliable automation of cloud management services, the modules were made available to customers.

In a study, the authors examined the optimal pricing techniques for a CSP, based on the assumption of user upgrading costs and switching costs (Nan, Zhang, & Li, 2019). According to their findings, there is no one-size-fits-all market structure in equilibrium. The willingness of customers to upgrade and switch will be influenced if the incumbent offers lower pricing to its existing customers and sets lower thresholds for them to upgrade and switch at a cost. When dealing with clients who vary in their sensitivity to relevant expenses, the incumbent firm may be compelled to offer discounts to new customers rather than those who have previously made purchases. One of the major challenges that an on-premises software firm may face is competition from SaaS vendors, particularly in acquiring new customers from the untapped market.

Cloud professionals and decision-makers can enhance their ability to make accurate judgments by utilizing rough sets to model and evaluate uncertainty (Radulescu, Radulescu, Boncea, & Mitan, 2018). This method combines two multi-attribute decision-making techniques, namely the analytic hierarchy process (AHP) and simple additive weight (SAW). The AHP rough technique is employed to determine the subjective weights associated with each criterion, as described in the research. To address uncertainty in CSP evaluations, a basic SAW method is utilized. In this case study, a group of cloud experts were engaged to provide their perspectives.

7.3 CLOUD PLATFORM

The cloud provider makes resources available to the end user and manages them on their behalf, rather than the cloud provider themselves owning and managing the resources. Not so long ago, companies and individuals who used computers were required to purchase their own software and hardware and be responsible for their upkeep in order to make use of them. Applications, memory, programs, and gadgets hosted in the cloud have made it possible for people and organizations to access a wide range of computer resources on demand. It is no longer essential for users of cloud computing to invest the time, expense, and expertise required to acquire and maintain computing resources if they transition from on-premises to cloud-based resources for their data storage and processing needs. This unprecedented access to computing resources has led to the emergence of new businesses in various fields. People can now collaborate with colleagues through video conferencing and other collaborative platforms, access entertainment and educational content on demand, interact with home devices, request a taxi, and even rent vacation accommodations in someone's home. All of these possibilities are made possible by the cloud.

7.3.1 CHARACTERISTICS

The following features may be found in cloud computing:

1. **Resources Pooling:** An architecture with multiple tenants has been implemented by the Cloud provider to enable accessibility of computer resources to a large number of customers. Different physical and virtual resources are allocated and reallocated based on the requirements of the client base. The allocation of resources is typically beyond the control of the client; however, they can be described at a higher level of abstraction if desired.

2. **Self-Service on the Go:** It is a critical consideration and helpful characteristic of cloud computing because it is able to continuously monitor server uptime, capabilities, and assigned network storage, among other things. This feature also enables the user to keep an eye on the overall performance of the machine.
3. **Simple to Maintain:** In many cases, there is no downtime, and the servers can be easily controlled by administrators. Each new edition of Cloud Computing that is released enhances the technology even further compared to its predecessor. Recent updates bring several advantages, such as improved interoperability with devices, bug fixes, and other enhancements.
4. **Large-Scale Networking:** To access or upload data to the cloud, all that is necessary is a device and an internet connection on the user's end. These characteristics may be found all across the internet as a result of the employment of the technology.
5. **High Availability:** The capabilities of the Cloud can be further enhanced and customized to meet the specific requirements of each individual user. Customers who require additional storage space can obtain it at a very low cost through cloud storage providers.
6. **Secure:** Another highly advantageous aspect of cloud computing is the level of protection it provides. It creates backups of data, ensuring accessibility even if one of the servers goes down. The data is stored securely on devices that cannot be hacked or accessed by unauthorized individuals. The storage facility is renowned for its quick and reliable service to customers.
7. **Pay as You Go:** When it comes to cloud computing, consumers only pay for the services that they truly utilize and consume. There are no hidden fees or additional charges to worry about. It is a cost-effective solution that often includes free storage space.

7.3.2 CLOUD COMPUTING CLASSIFICATIONS

Remember that not all clouds are the same, and not all cloud computing options are suited for every case or user. In order to assist us in making the best decision for our unique situation, a wide range of models, varieties, and services are offered. Choosing a cloud deployment or cloud computing design is the initial step before moving on to the next step (What Is Cloud

Computing? n.d.). Public, private, and hybrid cloud services may all be implemented on these three types of clouds (Figure 7.1):

1. **Public Cloud:** It is a computer resource that enables users to access computing resources such as servers and storage via the Internet through a third-party CSP. An increasing number of individuals are utilizing public clouds such as Microsoft Azure, Amazon Web Services (AWS), and others. Similar to private clouds, in this approach, the CSP is responsible for owning and managing the hardware, software, and other infrastructure. To utilize these services and monitor their account, a web browser is required.

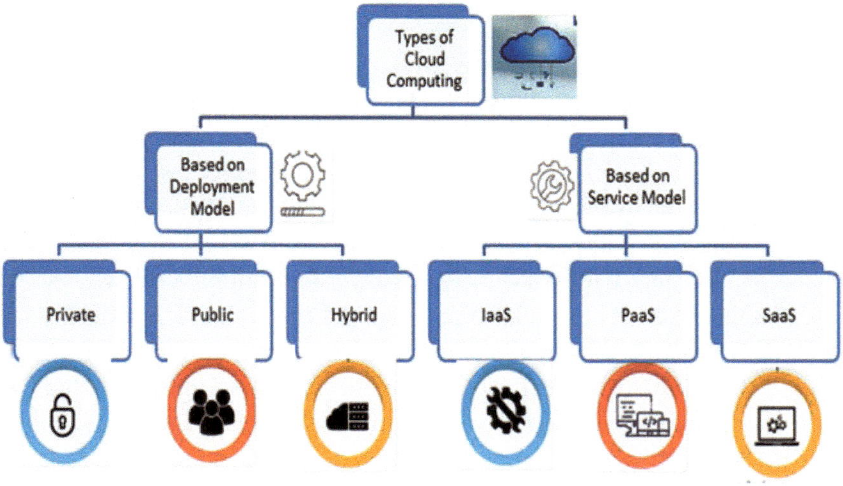

FIGURE 7.1 Types of cloud computing.

2. **Private Cloud:** It is common to use the phrase "private cloud" to refer to cloud computing capabilities that are only utilized by one firm or organization. For example, a private cloud may be located in a company's own data center, allowing for more flexibility. In certain cases, corporations contract out the provision of private cloud hosting to third-party service providers. Unlike public clouds, private clouds are located on a private network and provide services and infrastructure to users.
3. **Hybrid Cloud:** Hybrid clouds combine the best features of both public and private clouds, allowing for seamless data transfer between

the two. They offer enhanced flexibility, a wider range of deployment options, and improved security and compliance by enabling the flow of data and applications between private and public clouds.

7.3.3 TYPES OF CLOUD SERVICES

In cloud computing, there are four primary types of services: infrastructure as a service (IaaS), platform as a service (PaaS), serverless computing, and software as a service (SaaS). The term "cloud computing stack" refers to the arrangement of these components that is often used to refer to them. Understanding the distinctions and similarities between them will make it easier for developers to achieve their company's goals (Figure 7.2):

1. **Infrastructure as a Service (IaaS):** When needed, cloud computing services like IaaS supply critical computational and storage resources on-demand, according to a fee schedule. The other three categories of cloud services are SaaS, PaaS, and serverless computing. In addition to saving money on hardware maintenance, moving the company's infrastructure to an IaaS environment may also provide customers with real-time business information. When we use IaaS solutions, it is simple to scale up and down the IT resources. Aside from that, they make it simpler to deploy new applications and increase the general reliability of the system. In the case of IaaS, one can avoid the high costs and difficulties of purchasing, maintaining, and managing datacenter facilities, as well as our own servers. Each service is provided with a unique service component, and customers only pay for the resources that they utilize.
2. **Platform as a Service (PaaS):** Everything we need to build and deploy cloud-based applications, from basic apps to complex, cloud-enabled corporate systems, can be done from a single place and with little effort. As long as we have access to a secure Internet connection, CSPs will offer users the resources they need. A PaaS service, like an IaaS service, includes infrastructure—servers, storage, and networking—as well as middleware, development tools, business intelligence (BI) and database management systems, and other services. When it comes to web application creation and maintenance, there are many benefits to adopting a PaaS platform. By using PaaS, it is feasible to ignore the cost and complexity of

obtaining, installing, and managing software licenses, as well as middleware, container orchestrators such as Kubernetes, and other resources. Every other facet of the company's operations is often.

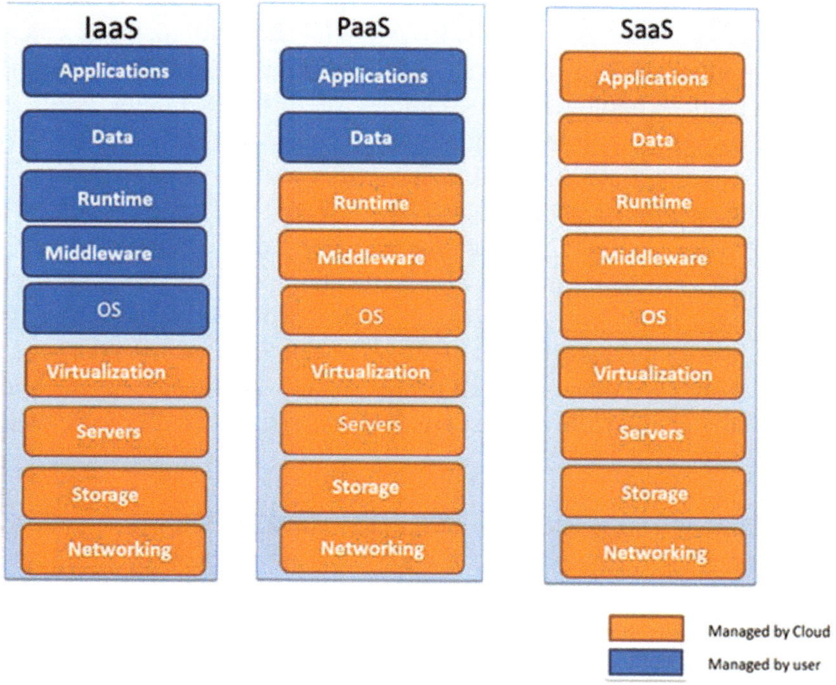

FIGURE 7.2 Comparing the types of clouds.

3. **Serverless Computing:** This accelerates the development handled by a third-party cloud computing service provider. New applications are created by relieving developers of the responsibility for infrastructure maintenance. The CSP sets up, scales, and maintains the infrastructure required to run the code in serverless applications on the cloud. It is critical to understand that the code is still being run on servers in order to fully appreciate serverless computing. A serverless application is one in which the developer does not have to be concerned with maintaining and supplying the infrastructure for their application, as the phrase suggests. This method allows developers to focus on business logic and provide more value to the company's core function as a consequence of their improved

ability. Serverless computing helps teams work more effectively and bring new ideas to market more quickly, while also helping organizations better manage their resources and focus on innovation through the application of cloud computing.

4. **Software as a Service (SaaS):** Customers can access and utilize cloud-based applications from any location with network connectivity when using SaaS. Email, calendars, and other commonly used technologies are examples of this. Essentially, SaaS is a comprehensive software solution that is obtained from a CSP on a pay-per-use basis. Business users are provided access to the application through a web browser and are charged a monthly subscription fee. The service provider's data center houses all the infrastructure, middleware, application software, and application data. If we have a service agreement with the provider, they will handle the hardware, software, as well as the application and its associated data on behalf of the users. SaaS enables organizations to quickly deploy and operate applications at an affordable price.

7.4 COMPARE CLOUD SERVICE PROVIDERS (CSPS)

When trying to develop a method for evaluating CSPs, it can be challenging. Many providers may appear similar at first, but upon closer inspection, they are actually distinct entities. To put it simply, comparing Cloud service companies is a difficult task.

Given the rapid expansion in cloud computing, service providers have been quick to reinvent their cloud products, creating new cloud offerings, price levels, and other ways to distinguish their offerings from their competition. As a result, it has become increasingly difficult to do an exact pricing comparison in the marketplace. It may be tough to evaluate and contrast cloud computing suppliers and the services they provide.

7.4.1 CLOUD COMPUTING INDUSTRY CHALLENGES

In addition to conducting due diligence on vendors, procurement professionals should also consider the fundamental characteristics of Cloud services. There are various types of "cloud computing" services available, which rely on a provider's physical servers that are partitioned using virtualization methods and then offered as "IaaS compute service offerings" to customers. To make an accurate comparison, it is necessary to have the following information:

- Bandwidth availability and network;
- A server's internal components, such as its processor, RAM, and storage;
- Type and implementation of virtualization technology (or hypervisor).

7.4.2 THE ADVANTAGES OF CLOUD TRANSPARENCY

Cloud transparency can bring benefits to various stakeholders, including end-user corporations, individual users, and CSPs. Having transparency in the cloud is advantageous for several reasons, some of which are listed below:

- Cloud transparency enables end-user organizations to compare and analyze service providers as well as cloud services in order to choose the best option for their specific applications.
- Facilitating companies can match their individual workloads to suitable cloud services using cloud transparency, a method that combines performance data with pricing data and price-performance analytics. This makes it easier to migrate workloads to the cloud.
- By utilizing price-performance analytics, cloud transparency data can be leveraged to optimize cloud cost and cloud portfolio management operations. Data-driven approaches to determine the highest value in cloud computing pave the way for efficient long-term cost reduction by assisting organizations in migrating workloads to the most cost-effective provider and service.
- It is possible to simulate application workloads and anticipate yearly operational costs for hosting them on several CSPs' services using proper modeling and simulation tools. Workload forecasting and simulation are now possible. There may be a surge in cloud arbitrage, which is the process of choosing the best cloud services given the current pricing and performance factors.
- This allows for price-based cloud provider arbitrage and analytics based on performance. It will ultimately be able to achieve the ultimate aim of cloud management: transferring workloads across cloud providers based on their actual performance and cost. The

cloud transparency data must be easily accessible in the many cloud management tools that are now available in order for this to operate.
- To establish a genuine industry standard for comparing the pricing and performance of different vendors, there should be an industry-wide perspective of price and performance data provided by cloud transparency. This data can be utilized by CSPs and their specialized cloud solutions.

Amazon, Google, and Microsoft are the world's three major CSPs, with a few smaller regional companies providing services in specific regions of the world.

Making a decision on which of the "Big Three" (if any) to concentrate on is an important first step. Adding a smaller service provider that offers unique services and value to the company may be something we want to consider in certain situations. Additionally, aspects such as simplicity of use, self-service exposure to a company's services, and a comprehensive API that provides easy access to the offered Cloud services are crucial.

7.4.3 COMPARISON BASED ON TYPE OF CLOUD PROVIDER

Users depend on service providers, therefore they should include not just the capacity to assess their firm's stability, dependability, and overall capabilities and pricing, but also a thorough understanding of the numerous underlying components of Cloud services, such as the following:

- Regions and countries in which service providers operate;
- The locations of data centers;
- Scale and long-term viability of a company;
- Service providers reliability;
- Transparency in audit and compliance;
- Contract conditions and pricing models;
- Agreements regarding the level of service;
- Self-service, a rich API, and ease of usage.

7.4.4 COMPARISON BASED ON SERVICES

It is not as straightforward as one would expect to evaluate the various cloud services offered by different CSPs. The following are some considerations to bear in mind while considering service offerings:

- Specifications for a certain product or service.
- Services offered by a cloud computing platform include virtual machines (VMs) and VM families (VMs). These resources are often measured in terms of the ratio of virtual CPUs to virtual RAM (vRAM).
- Disk-based vs solid-state storage; network-attached vs local; spinning vs solid-state (SSD).
- Generation of CPUs and their technologies.
- An objective benchmarking procedure is used to measure actual performance.
- The technology that underpins it (those not shared with end-users).
- Additional services, such as support, managed services, and add-ons.

The attributes of the cloud service on the provider's website are usually used as a starting point for comparison by end-users, followed by a price comparison. When running workloads, it's crucial to remember that the specifications of any given virtual machine are just an estimate of how well it will perform in practice.

If we consider the total number of virtual CPUs, physical server chip technology, RAM size and type, as well as storage type and volume size, these are all factors that can be studied quickly and efficiently. While providers often highlight the specifications on their websites, the truth is that performance is not solely determined by these specifications. It is crucial to delve deeper into the pricing details to make a more realistic comparison.

7.4.5 COMPARISON ON THE BASIS OF PRICES

These days, finding out how much a Cloud service will cost for the user is a mystery. In order to stay competitive, cloud service companies regularly provide updated versions of their existing services, as well as introduce new services and pricing schemes. It is quite uncommon for pricing systems to be more complex than others. To accurately assess the cost of a Cloud service, it is necessary to consider the following factors, both for a single service and when multiple Cloud services are combined for a single business use case.

- Computing services, classes, and instances, storage solutions and choices, network services, etc., are just a few examples of the many services that may be priced separately.

- Increments in billing.
- Unused capacity may be sold at a discount in order to achieve the goal of asset utilization through the usage of reserved compute services and up-front concessions in comparison to spot instances.
- Strategies for long-term discounts.
- Services can be purchased as a package instead of individually, such as a local SSD and a VM. Storage is included in certain VM packages, while it is not included in others.
- For multinational corporations, currency disparities and swings are an issue.

Price cuts are common, and savings can be achieved for long-term contracts with a single supplier. The massively scalable providers are engaged in a "race to the bottom" to gain market share, which will eventually reach a point of stagnation. For now, customers will benefit from this development.

Additionally, cloud deployments require more than one server, storage, and connectivity, as well as other services, to meet the complete application demands. Cloud computing is becoming more affordable, even though a corporation might benefit from discounts on reserved compute resources if it commits for a longer period of time. However, it may be counterproductive to sign up for a long-term contract based on a current discount if prices fall in the future. Consumers may need a team of financial modelers to understand the alternatives and scenarios available to them when considering cloud competitive pricing and scenarios. For example, users should be able to monitor the costs of their cloud service providers (CSPs) and make well-informed choices about which companies to work with and which cloud services to take advantage of.

7.5 CONCLUSION

This research mainly aims to investigate cloud computing platforms, which are a great resource since they contain so much data. More the many kinds of clouds and the wide variety of services they provide are covered in this one easy-to-navigate spot, so you can learn all about cloud computing. The research goes deep into analyzing the pros and cons of various cloud service providers by comparing and contrasting them. The research intends to guarantee a complete comprehension of the whole scene by offering a detailed analysis of each tactic. In order to make an educated comparison, factors including the service provider's kind, the services supplied, and pricing structures are carefully examined.

KEYWORDS

- **Amazon web services**
- **cloud platform**
- **cloud service provider**
- **cloud transparency**
- **infrastructure as a service**
- **platform as a service**
- **serverless**
- **software as a service**

REFERENCES

Arockiam, L., Monikandan, S., & Parthasarathy, G., (2011). Cloud computing: A survey. *International Journal of Internet Computing, 1*(2), 26–33.

Baron, J., & Kotecha, S., (2013). *Storage Options in the AWS Cloud*. Amazon Web Services, Washington DC, Tech. Rep.

Beri, R., & Behal, V., (2015). Cloud computing: A survey on cloud computing. *International Journal of Computer Applications, 111*(16).

Chand, M., (2021). *Top 10 Cloud Service Providers IN 2021*. https://www.c-sharpcorner.com/article/top-10-cloud-service-providers/ (accessed on 11 January 2023).

Chappell, D., (2008). *Introducing the Azure Services Platform* (Vol. 1364, No. 11). White paper.

Chappell, D., (2009). *Introducing Windows Azure*. Microsoft, Inc, Tech. Rep.

Chappell, D., (2010). *Introducing the Windows Azure Platform*. David Chappell & Associates White Paper.

Copeland, M., Soh, J., Puca, A., Manning, M., & Gollob, D., (2015). *Microsoft Azure*. New York, NY, USA: A press.

Dharmapriya, W. A. S. P., Supipi, K. G., Nimesh, G. R., Muhandiram, M. A. B. K., Rankothge, W. H., & Gamage, N., (2019). Smart platform for cloud service providers. In: *2019 International Conference on Advancements in Computing (ICAC)* (pp. 434–439). IEEE.

Dubey, P., Chhabra, G. S., Bhattacharya, R., & Singh, T. G., (2021). Enhancing the performance of traditional Hadoop using AWS with EMR approach. *Design Engineering*, 571–580.

Fernando, N., Loke, S. W., & Rahayu, W., (2013). Mobile cloud computing: A survey. *Future Generation Computer Systems, 29*(1), 84–106.

Five Challenges to Adopting AWS–the Bunker, (2017). The Bunker. https://www.thebunker.net/blog/cloud-infrastructure/5-challenges-to-adopting-aws/ (accessed on 11 January 2023).

Garraghan, P., Townend, P., & Xu, J., (2013). An analysis of the server characteristics and resource utilization in google cloud. In: *2013 IEEE International Conference on Cloud Engineering (IC2E)* (pp. 124–131). IEEE.

Gartner Recognizes AWS As a Magic Quadrant Leader for Cloud Infrastructure & Platform Services. (n.d.). *Amazon Web Services, Inc.* https://aws.amazon.com/resources/analyst-reports/gartner-mq-cips-2021/?pg=WIAWS-mp (accessed on 11 January 2023).

Ghosal, I., (2015). A demographic study of buying spontaneity on e-shoppers: Preference Kolkata (West Bengal). *Journal of Technology Management for Growing Economies, 6*(2), 65–75.

Islam, M. M., Morshed, S., & Goswami, P., (2013). Cloud computing: A survey on its limitations and potential solutions. *International Journal of Computer Science Issues (IJCSI), 10*(4), 159.

Jaithunbi, A. K., Sabena, S., & SaiRamesh, L., (2021). Trust evaluation of public cloud service providers using genetic algorithm with intelligent rules. *Wireless Personal Communications, 121*(4), 3281–3295.

Jiang, C., Han, G., Lin, J., Jia, G., Shi, W., & Wan, J., (2019). Characteristics of co-allocated online services and batch jobs in internet data centers: A case study from Alibaba cloud. *IEEE Access, 7*, 22495–22508.

Kochut, A., Deng, Y., Head, M. R., Munson, J., Sailer, A., Shaikh, H., & Wagner, H., (2011). Evolution of the IBM Cloud: Enabling an enterprise cloud services ecosystem. *IBM Journal of Research and Development, 55*(6), 7:1–7:13.

Li, K., (2021). On the profits of competing cloud service providers: A game theoretic approach. *Journal of Computer and System Sciences, 117*, 130–153.

Marks, B. E., (2018). How to compare cloud providers, services, and prices in 2020 | Cloud spectator. *Understanding Heart Disease* (1992). http://books.google.ie/books?id=ypm3DAEACAAJ&dq=Understanding+heart+disease&hl=&cd=1&source=gbs_api (accessed on 11 January 2023).

Moreno, I. S., Garraghan, P., Townend, P., & Xu, J., (2013). An approach for characterizing workloads in google cloud to derive realistic resource utilization models. In: *2013 IEEE Seventh International Symposium on Service-Oriented System Engineering* (pp. 49–60). IEEE.

Mukherjee, S., (2019). *Benefits of AWS in Modern Cloud.* Available at SSRN 3415956.

Nan, G., Zhang, Z., & Li, M., (2019). Optimal pricing for cloud service providers in a competitive setting. *International Journal of Production Research, 57*(20), 6278–6291.

Parkhomenko, A., Zalyubovskiy, Y., & Parkhomenko, A., (2021). Cloud platform for software development. In: *International Conference on Remote Engineering and Virtual Instrumentation* (pp. 343–351). Springer, Cham.

Prasad, B., & Ghosal, I., (2021). Forecasting buying intention through artificial neural network: An algorithmic solution on direct-to-consumer brands. *FIIB Business Review*, 23197145211046126.

Rădulescu, C. Z., Rădulescu, I. C., Boncea, R., & Mitan, E., (2018). A group decision approach based on rough multi-attribute methods for cloud services provider selection. In:

2018 10th International Conference on Electronics, Computers and Artificial Intelligence (ECAI) (pp. 1–6). IEEE.

Rahimi, M. R., Ren, J., Liu, C. H., Vasilakos, A. V., & Venkatasubramanian, N., (2014). Mobile cloud computing: A survey, state of art and future directions. *Mobile Networks and Applications, 19*(2), 133–143.

Ramamurthy, A., Saurabh, S., Gharote, M., & Lodha, S., (2020). Selection of cloud service providers for hosting web applications in a multi-cloud environment. In: *2020 IEEE International Conference on Services Computing (SCC)* (pp. 202–209). IEEE.

Sampé, J., Vernik, G., Sánchez-Artigas, M., & García-López, P., (2018). Serverless data analytics in the IBM cloud. In: *Proceedings of the 19th International Middleware Conference Industry* (pp. 1–8).

Trueman, T. E., Narayanasamy, P., & Ashok, K. J., (2021). A graph-based method for ranking of cloud service providers. *The Journal of Supercomputing*, 1–18.

Varia, J., (2010). Migrating your existing applications to the AWS cloud. *A Phase-driven Approach to Cloud Migration*, 1–23.

Varia, J., (2011). Best practices in architecting cloud applications in the AWS cloud. *Cloud Computing: Principles and Paradigms, 18,* 459–490.

What Is Cloud Computing? A Beginner's Guide | Microsoft Azure. (n.d.). *What is Cloud Computing? A Beginner's Guide | Microsoft Azure*. https://azure.microsoft.com/en-in/overview/what-is-cloud-computing/#cloud-deployment-types (accessed on 11 January 2023).

Zhan, C., Su, M., Wei, C., Peng, X., Lin, L., Wang, S., & Chai, C., (2019). Analytic dB: Real-time OLAP database system at Alibaba cloud. *Proceedings of the VLDB Endowment, 12*(12), 2059–2070.

Zhang, G., & Ravishankar, M. N., (2019). Exploring vendor capabilities in the cloud environment: A case study of Alibaba cloud computing. *Information & Management, 56*(3), 343–355.

Zhu, J., Fang, X., Guo, Z., Niu, M. H., Cao, F., Yue, S., & Liu, Q. Y., (2009). IBM cloud computing powering a smarter planet. In: *IEEE International Conference on Cloud Computing* (pp. 621–625). Springer, Berlin, Heidelberg.

CHAPTER 8

India: The Digital Economy and Its Financial Inclusion with Special Reference to Industry 5.0

PIYALI ROY CHOWDHURY[1] and INDRAJIT GHOSAL[2]

[1]*Institute of Management Study, West Bengal, India*

[2]*Associate Professor, School of Management and Commerce, Brainware University, Barasat, Kolkata, West Bengal, India*

ABSTRACT

Digitalization of the Indian economy is crucial in the current economic scenario. In the recent past, online portals for financial transactions have also gained popularity. Prior to the deregulation of the Indian economy, digitization in financial activities was limited to dematerialized accounts for stock trading. Real-time gross settlement (RTGS) and National Electronic Fund Transfer (NEFT) in banking have brought about a huge revolution in the financial sector. By permitting bank account users to continue using internet banking, the enormous conduit for electronic fund transfer reached the next level. In light of this financial market upheaval, the current chapter examines the relationship between the digital economy and financial inclusion in India from 2005 to 2020. Prior to 2005, India's use of digital payments was deemed insignificant. Using the autoregressive distributed lag (ARDL) modeling approach, a cointegration between the number of automated teller machines (ATMs) and the number of mobile subscriptions in India is determined. The data collected for the investigation conform to time series data flow; hence, econometric modeling is preferable. The

Sustainable Strategic Business Infrastructure Development and Contemporary Digital Practices in Industry 5.0. Indrajit Ghosal, Bikram Prasad, Nilanjan Ray, & Lalit Kumar Joshi (Eds.)

information was extracted from the annual database of the World Bank between 2005 and 2020. The purpose of this study is to determine the effect of rural and urban mobile subscriptions on India's ATMs. The research identifies a strong cointegration between the two variables. It is discovered in the short term that the number of ATMs in a country decreases as the number of mobile subscribers increases. There is a 67% possibility that the short-term shocks will be eliminated and replaced by long-term stability. In conclusion, the significance of the error correction term indicates the existence of a long-term effect of mobile subscriptions on ATMs. The Jarque-Bera statistics indicate that the data in the series are normally distributed. The report reveals the significance of mobile usage and digital payments in India, as well as the gradual decline of conventional ATMs in India. In addition, the necessity of financial inclusion has been addressed throughout the research. The requirement for mobile subscriptions leads to smart payments and electronic financial transactions regardless of the country's sector-specificity. The study also improves the utilization of network coverage in terms of Internet connectivity in India's urban and rural areas.

8.1 INTRODUCTION

In the recent past, India has undergone significant changes in its payment system, which has led to the increased adoption of digital technology. The application of digital technology has revolutionized the financial market in various ways. To facilitate the transformation of digital transactions in the financial market, there is a need for improved connectivity among people through enhanced network infrastructure and better technology, particularly in terms of mobile connectivity. In this respect, the importance of internet of things (IoT) with respect to mobile financing has been increased to a great extent. India is gradually trying to become a cashless economy. However, in some parts of the country, digitalization is still yet to thrive.

According to the Boston Consulting Group, countries with a high concentration of cash tend to have less economic growth compared to cashless economies. Therefore, for India to further its growth, it is essential to transition towards cashless financial transactions. One of the advantages of moving towards cashless transactions is that it is not limited by specific time constraints. With strong and reliable network connectivity, it enables promptness and accuracy in every transaction, regardless of the transaction volume.

The concentration of the urban and rural population also contributes to the movement of cashless initiatives. Hence, the importance of financial inclusion is considered a top priority in India. However, the basic knowledge of the financial market is relatively poor, resulting in a lower number of bank accounts in the Indian market. This leads to a lack of accountability for the holders of the total cash volume in specific areas of the country. The transition from cash-based to cashless transactions has already begun in different parts of India during and after the COVID-19 pandemic. The need for a cashless economy has also been well understood in terms of integrating all financial applications during the same period. The initiatives taken by the Government of India are highly appreciated, regardless of the financial volume in the country.

The role of smartphone penetration, effective network connections, ease of doing business due to Know Your Customer (KYC) relaxation, private banks-led payment applications such as Pockets, PayZapp, and government-led applications has made business transactions easier, irrespective of the geographical boundaries of India. It proves that the creativity in human minds, with reference to technology and machinery usage, has resulted in efficiency in resource usage and user-friendly interfaces in financial transactions. In this aspect, the effectiveness of digital transactions based on Industry 5.0 is a crucial point to be analyzed for the overall health of the Indian financial market in recent time periods. On the other hand, the strongest short-run external shock, Coronavirus, has compelled India to move towards a contactless digital economy through digital transactions across all sectors. The financial inclusion with respect to digital transformation in the Indian market, thus, needs to be reviewed and reformed in an effective way to create sustainability in the financial sector.

Financial services, along with the concept of cloud data gathering, are vital in the processing of financial data. This relates to the automation of financial data, which can also be highly related to Industry 5.0 in the financial sector. Cloud-based financial data can provide important information in the financial sector in the near future and can also set the trend for forecasting the overall health of the Indian financial market.

Based on the points mentioned above, the current study analyzes the role of smartphones in relation to rural and urban mobile subscriptions on India's automated teller machines (ATMs). The study has three main objectives: to determine the long-term and short-term impacts of rural and urban mobile subscriptions on India's ATMs in India, to find the long run directional causality among the variables and to find the short run granger causality with model stability in short as well as long run.

8.2 REVIEW OF LITERATURE

The current study analyzes the role of digitalization on financial inclusion in the Indian economy. There are few relevant studies in the related area of research. The relationship between these two variables is considered crucial in the near future. The level of development depends on the extent of digitalization of the country as a whole. The literature review is divided into two main aspects: the association between the digital economy and financial inclusion, and the role of Industry 5.0 in the digital economy to enhance financial inclusion.

8.2.1 ASSOCIATION BETWEEN DIGITAL ECONOMY AND FINANCIAL INCLUSION

Vyas & Jain (2021) explore the role of digitalization on the overall financial health of India. The study adopts factor analysis to prove that education is the key input for the digital economy. However, it fails to take a broad view of the effect of digitalization on the population of India. Nagpal et al. (2019) investigate the role of education, gender, government intervention, and level of earning in a family to determine the aggressiveness of financial inclusion in India. They focus on the technology usage measured as internet application in the areas by individuals. Agrawal & Jain (2019) find out the behavioral acceptance of retail banking customers towards new technology adoption in the financial market of India. They focus on analyzing the acceptability and usage convenience of digital transactions among banking customers. Singh et al. (2018) study the factors considered for the payment system in India using digital methods. They explore crucial factors such as acceptability, knowledge, and usage consistency as the main factors to promote digitalization in India. Sankar et al. (2018) find out the effect of demonetization on financial inclusion in India. It demonstrates a positive association between the two variables. Ghosal & Prasad (2019) explain the digitalization in the handicraft market in India. Additionally, Ghosal et al. (2020) explore the digitalization in the Indian rural and urban market. Siddik et al. (2020) investigate the issue of digitalization in developing economies. Chowdhury & Tiwari (2020) analyze the health of the financial market in India during the coronavirus period. They also investigate the status of the Indian stock market by considering the economic growth in India. The focus is on strategies to promote digital financial services in rural regions with poor financial status. Ni (2020) proves that acceptability, perception, and ease of use influence digital finance in developing economies like Malaysia. Wibella et al. (2018) analyze the role of digitalization in Bogor city, Jakarta. The study confirms that perception is one of the important factors in stimulating digitalization in the financial sector in developing economies. Tiwari et al. (2019) explore the difficulties of implementing digitalization in the financial sector in India. They identify issues related to income level, knowledge, and education that exacerbate financial inclusion in developing economies. Rastogi & Ragabiruntha (2018) prove that knowledge of banking and online financial transactions will improve the overall financial structure in the developing economy.

8.2.2 ROLE OF INDUSTRY 5.0 IN THE DIGITAL ECONOMY TO AGGRAVATE FINANCIAL INCLUSION

Carayannis & Morawska-Jancelewicz (2022) analyze the role of Industry 5.0 on the future economy. They examine the implementation of automation in developing economies and propose policy measures to ensure economic sustainability through technological advancement. Cillo et al. (2021) explore the concept of social development through automation in industries within developing economies. They find that cultural progress through automation can significantly promote economic growth. Xu et al. (2021), Alexa et al. (2022), and Saniuk et al. (2022) focus on the differences between Industry 4.0 and 5.0, as well as their impact on the economy. Industry 5.0 is an emerging field of study, with researchers concentrating on various areas and sectors that will contribute to the overall development of the economy.

Analyzing the literature reviews mentioned above, the current study identifies several gaps. The previous studies primarily focus on the overall findings of digitalization, without utilizing specific variables to analyze the concept of financial inclusion. Additionally, no econometric models have been employed to examine the issue, and the use of secondary data from databases for statistical analysis has not been considered before. Previous studies are predominantly theoretical or empirical in nature, utilizing primary data sources and considering qualitative factors. In contrast, the current study employs quantitative variables to analyze an econometric model. It is also unique in its consideration of region-specific datasets, such as rural-urban data, while previous studies are based on the overall country. Furthermore, this study specifically focuses on different sectors within the country. Lastly, the study explores the relationship between digital economy, financial inclusion, and Industry 5.0, which is a relatively new area of research. The study also discusses the applicability of automation in financial inclusion, leading to digitalization.

8.3 DATA DESCRIPTION AND ANALYSIS

This study examines the relationship between India's digital economy and financial inclusion. In this regard, proxy factors such as rural-urban mobile subscriptions and the number of ATMs are taken into account for measuring financial inclusion. The variables are described based on the information provided by the World Bank database. The data covers a period of 15 years, from 2005 to 2020. Mobile cellular phone subscriptions refer to public mobile phone services that connect to the PSTN using cellular technology, as defined by the World Bank. This statistic includes both the number of post-paid subscriptions and the number of active prepaid accounts (i.e., those that have been used during the last three months). The indicator is compatible with any mobile cell phone plans that permit phone calls. It excludes subscriptions to public mobile data services, private trunked mobile radio, telepoint, radio paging, and telemetry services, as well as data card and USB modem subscriptions. Additionally, the definition of ATM as world bank can be stated as follows: ATMs are computerized telecommunications equipment that enable bank customers to conduct financial transactions in public.

The hypothesis of the current study has been explained as follows:

- **H_{01}:** There is no cointegration between rural-urban mobile subscriptions and the number of automated teller machines (ATMs) in India.
- **H_{02}:** There is no impact of rural-urban mobile subscriptions and the number of automated teller machines (ATMs) in India.
- **H_{03}:** There is Granger causality among rural-urban mobile subscriptions and the number of automated teller machines (ATMs) in India.

The study considers the autoregressive distributed lag modeling approach (ARDL) to analyze the data. The advantage of ARDL is that it can be applied irrespective of the data integration. Even if the data is integrated to different levels, the outcomes cannot be spurious. Additionally, it can provide both a long-run and short-run overview of the data. Therefore, the significance of the variables can be well formulated with the application and implication of the variables in the Indian economy.

Considering two variables, rural-urban mobile subscriptions and number of ATMs, the study contemplates econometric analysis. Since the data is in two different formats, the logarithmic transformation of the data has been considered. Initially, unit root analysis was conducted to determine the integration of the data. The results of the unit root analysis, which analyze the stationarity of the data, are presented in Table 8.1.

TABLE 8.1 Result of Augmented Dicky Fuller (ADF) and Phillips Perron (PP) Unit Root Test

Variables	I (0)		I (1)	
	ADF	PP	ADF	PP
l (Rural-urban mobile subscriptions)	(4.49)**	(4.49)**	–	–
l (Automated teller machines (ATMs))	(3.34)**	(2.68)	–	(6.09)**

**Signifies the data is significant at 5% alpha level.

Table 8.1 demonstrates that both the variables, rural-urban mobile subscriptions and number of ATMs, are of I(0) nature. This indicates that the data are stationary at their levels.

Table 8.2 indicates the presence of long-run cointegration between rural-urban mobile subscriptions and the number of ATMs. It demonstrates the long-term impact of mobile subscriptions on ATMs in the Indian economy. The coefficient of the dependency, however, can be determined from the short-run error correction model in Table 8.3.

TABLE 8.2 Result of Auto Regressive Distributed Lag Modeling Approach (ARDL) Bound Test

Variables		F Statistic	I (0)		I (1)	
Dependent Variable	Independent Variable		5%	10%	5%	10%
l (Automated teller machines (ATMs))	l (Rural-urban mobile subscriptions)	11.55**	3.62	3.02	4.16	3.51

**Signifies the data is significant at 5% alpha level.

TABLE 8.3 Result of Short Run Error Correction Model (Dependent Variable – ATMs; and Independent Variable – Mobile Subscription)

Variables	Coefficients	T Statistic	Probability
l (Automated teller machines (ATMs))_1	(0.18)	(1.51)	0.27
l (Automated teller machines (ATMs))_2	(0.33)*	3.23	0.08
l (Automated teller machines (ATMs))_3	0.13	1.33	0.31
l (Rural-urban mobile subscriptions)	(0.16)	(1.24)	0.34
l (Rural-urban mobile subscriptions)_1	0.13	0.75	0.53
l (Rural-urban mobile subscriptions)_2	(0.76)**	(4.92)	0.04
l (Rural-urban mobile subscriptions)_3	(0.43)*	(3.48)	0.07
Error correction term (ECT)	(0.67)	(8.32)	0.01

R-squared	0.98	Mean dependent var	0.06
Adjusted R-squared	0.95	S.D. dependent var	0.05
S.E. of regression	0.01	Akaike info criterion	(5.86)
Sum squared residual	0.00	Schwarz criterion	(5.53)
Log likelihood	43.13	Hannan-Quinn criterion.	(5.98)
Durbin-Watson	2.6		

*Signifies the data is significant at 10% alpha level.

**Signifies the data is significant at 5% alpha level.

Table 8.3 analyzes the short-run framework of the model, proving the short-run significance of the coefficients. The results demonstrate that higher mobile subscriptions will reduce the usage of ATMs in the country. The analysis also examines the possibility of the model transitioning from short-run disturbances to long-run stability, indicating that the model has a 67% capability to move towards long-run equilibrium. Lastly, the lag values of the independent variables highlight the importance of past periods in influencing the present economic situation.

Finally, it explains the short-run Granger causality through the Granger-Causality Test. The result of the Granger causality is shown in Table 8.4.

TABLE 8.4 Result of Granger Causality Test

Variable Description	F Statistic	Probability
l(Rural-urban mobile subscriptions) does not granger cause l (automated teller machines (ATMs)).	1.52	0.26
l(Automated teller machines (ATMs)) does not granger cause l(Rural-urban mobile subscriptions).	0.07	0.92

Table 8.4 explains the short-run Granger causality between the two variables. In addition to the short-run error correction model, the result of Granger causality helps identify the directional causality in the short run. The result proves that there is no bidirectional causality between rural-urban mobile subscriptions and ATMs.

The consistency of the model is analyzed through the result of the variance inflation factor (VIF). It is described in Table 8.5.

TABLE 8.5 Result of Variance Inflation Factor (VIF)

Variable	Coefficient	Centered VIF	Uncentered VIF
l(Automated teller machines (ATMs)) _1	0.076221	4363.986	207.2681
l(Automated teller machines (ATMs)) _2	0.079953	4135.250	284.5454
l(Automated teller machines (ATMs)) _3	0.069948	3197.540	300.3442
l(Automated teller machines (ATMs)) _4	0.037442	1475.403	178.6974
l(Rural-urban mobile subscriptions)	0.188628	687962.1	83.04784
l(Rural-urban mobile subscriptions) _1	0.203926	736694.5	213.3206
l(Rural-urban mobile subscriptions) _2	0.204238	728519.1	415.9169
l(Rural-urban mobile subscriptions) _3	0.213277	748939.0	700.7535
l(Rural-urban mobile subscriptions) _4	0.078552	270230.7	420.9522
C	4.628038	209441.3	NA

In Table 8.5, the VIF values can be found. Since all the values are less than 3, the model is proven to be consistent and significant for the implications of the study.

The autocorrelation, which measures the correlation among the variables and their lagged values, has been tested using the Lagrange-Multiplier (LM) test. The result indicates that there is no autocorrelation in the dataset (F value – 0.15 with a probability of 0.77). The heteroskedasticity, which determines the equality of variances among the residuals, has also been assessed using the Breusch Pagan Godfrey Test. The result shows that there is no heteroskedasticity in the dataset (F value – 0.13 with a probability of 0.99). Finally, the normality of the dataset has been evaluated using the Jarque-Bera (JB) test. The result suggests that the variables follow a normal distribution (JB – 4.70 with a probability of 0.10).

The stability of the overall model has been proven through cumulative sum (CUSUM) and cumulative sum of squares (CUSUMQ) tests. Both results confirm that the model is stable at a 5% level of significance. The results of CUSUM and CUSUMQ are depicted in Figures 8.1 and 8.2, respectively.

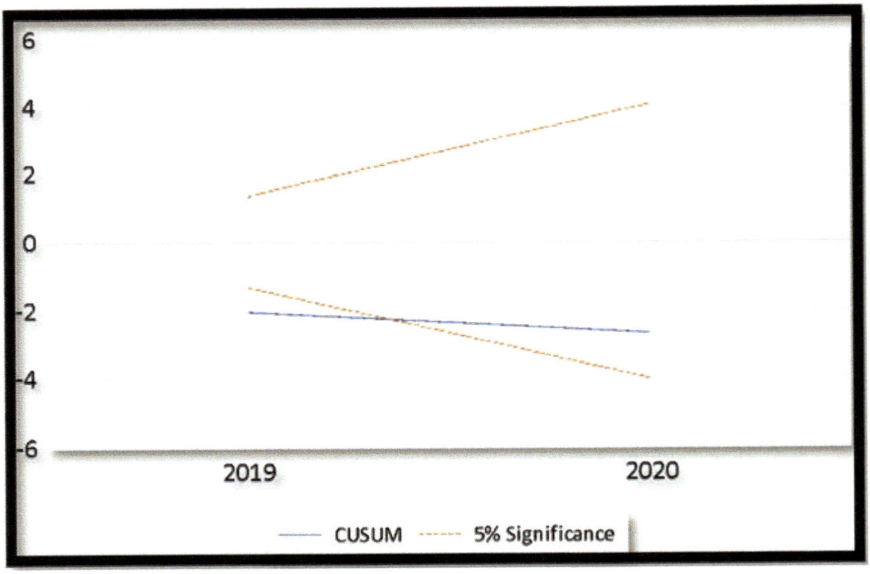

FIGURE 8.1 CUSUM test.

Figures 8.1 and 8.2 explain the stability of the model. It is proven that the model is stable at a 5% level of significance, as the blue line is placed between the upper and lower limits of the stability threshold.

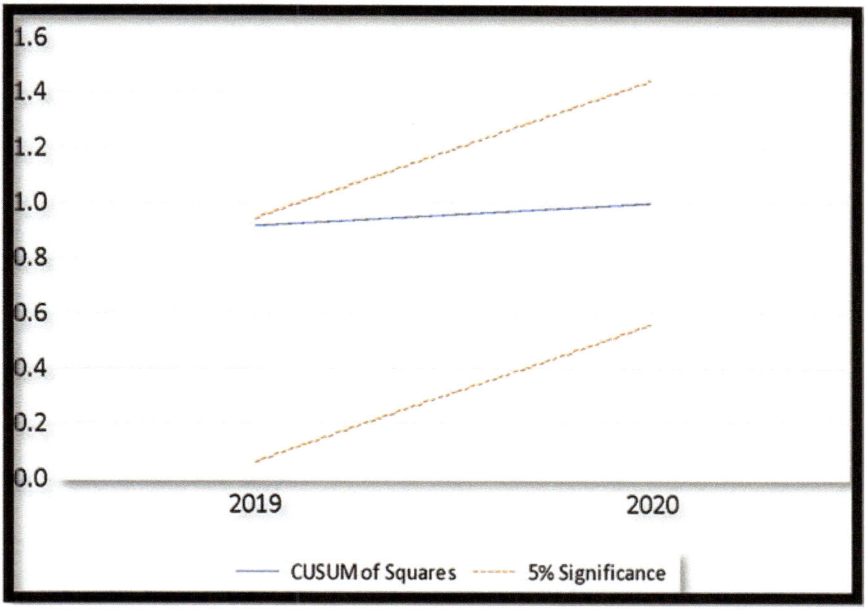

FIGURE 8.2 CUSUMQ test.

8.4 DATA INTERPRETATION AND IMPLICATION

The integration of data among the variables determines the choice of dataset methodology. Both variables considered in this study, rural-urban mobile subscriptions and ATMs, are integrated to I (0). Therefore, traditional econometric models may not yield optimal results for data analysis. On the other hand, ARDL is not restricted by data integration. Choosing ARDL allows for running the data without erroneous results and obtaining econometric regression results from the dataset. The result of the ARDL cointegration test indicates a significant cointegration between rural-urban mobile subscriptions and ATMs in the long run, with a significance level of 5% at 11.55. This implies that both variables move in the same direction in the long run, potentially providing future associations as they either increase or decrease together. This confirms the acceptance of H_{11}. Once the long run is established, ARDL examines short run phenomena in the model. The error correction model provides the short run coefficients of the model. In the study, it is found that an increase in mobile subscriptions in rural and urban areas leads to a reduction in the requirement of

ATMs in those areas. The study reveals that during the lag periods two and three of the rural-urban mobile subscription variable, the coefficients have significant values of 0.76 and 0.43 at a 10% level. This indicates that if mobile subscriptions increase by 1 unit, ATM establishment reduces by 0.76 and 0.43, respectively, over the past two and three periods. This confirms the acceptance of H_{12}. Additionally, the significant error correction term ensures the existence of long-run causality from rural-urban mobile subscription to ATMs in India. The value of the error correction term explains that there is a 67% chance for the Indian economy to correct itself from any short-run shocks and move towards long-run equality. The R square (98%) and the adjusted R square (95%) values demonstrate that the model is consistent in explaining the actual scenario. The Durbin Watson test value of 2.5 ensures negative autocorrelation in the dataset. Finally, the study proves that there is no short-run directional causality, as mentioned in the Granger causality, between the two variables. This confirms the acceptance of H_{03}.

The automation in the financial sector, specifically in retail banking, is a much-debated issue in a developing country like India. The success of automation can be seen in the service sector software industries or in the manufacturing sectors, but for the financial sector, it is a crucial aspect. Along with other macroeconomic factors, a developed economy is distinguished based on its financial components. Financial inclusion plays a bigger role in improving the economic status of a nation. India, being a developing economy, is engaging itself in more activities towards financial inclusion. Among these, the digitalization of the financial sector plays an important role in shaping the future financial health of the country. The implementation of huge mobile subscriptions for easing small payments as well as transactions will enhance the need for automation through Industry 5.0. The more digital the Indian economy becomes, the better the result of financial inclusion and the state of the economy.

8.5 CONCLUSION

The reality of the Indian economy lies in the high volume of cash transactions throughout the financial year. Apart from the service industry, such as software, retail-level transactions in the primary and secondary markets still heavily rely on cash operations. Additionally, there are noticeable differences in payment modes across the states of India. Therefore, addressing the sectoral and state-level payment imbalances, financial inclusion becomes highly essential. Simultaneously, it is crucial to prioritize the implementation of digitalization in payment methods. In this regard, the first step would be to enhance basic-level financial inclusion in the country. The focus should be on increasing the number of bank accounts, specifically targeting different regions and genders, as well as maintaining a minimum fund balance. The next level of emphasis should be on promoting digital literacy for retail transactions involving these accounts. Understanding the connection between bank accounts and payment applications is a significant advantage for the Indian economy to progress towards achieving success in Industry 5.0. Once individuals have a solid understanding of managing digital bank accounts, it will undoubtedly contribute to enhancing financial transactions across the country. This, in turn, will facilitate the smoother implementation of automation in the digital financial market in India.

KEYWORDS

- **automation**
- **data analysis**
- **data description**
- **data interpretation**
- **digital economy**
- **financial inclusion**
- **Industry 5.0**

REFERENCES

Agrawal, G., & Jain, P., (2019). Digital financial inclusion in India: A review. *Behavioral Finance and Decision-Making Models*, 195–203.

Alexa, L., Pîslaru, M., & Avasilcăi, S., (2022). From Industry 4.0 to Industry 5.0—An overview of European union enterprises. *Sustainability and Innovation in Manufacturing Enterprises*, 221–231.

Carayannis, E. G., & Morawska-Jancelewicz, J., (2022). The futures of Europe: Society 5.0 and Industry 5.0 as driving forces of future universities. *Journal of the Knowledge Economy*, 1–27.

Chowdhury, P. R., & Tiwari, N., (2020). Analyzing Indian stock market index volatility at the outbreak of corona virus: Can exchange rate be a determinant in measuring it? *Solid State Technology, 63*(5), 1503–1513.

Chowdhury, P. R., & Tiwari, N., (2020). *The Influence of Exchange Rate on Healthcare Stock Market Index in India, 83*, 24326–24331. ISSN: 0193-4120.

Cillo, V., Gregori, G. L., Daniele, L. M., Caputo, F., & Bitbol-Saba, N., (2021). Rethinking companies' culture through knowledge management lens during Industry 5.0 transition. *Journal of Knowledge Management.*

Ghosal, I., & Prasad, B., (2019). Inspiring digitalization of handicraft market: An empirical approach. *Parikalpana: KIIT Journal of Management, 15*(1, 2), 199–209.

Ghosal, I., Prasad, B., & Behera, M. P., (2020). Delineating the exchange environment of handicraft Industry from market space to marketplace: An inclusive map for RURBAN development. *Paradigm, 24*(2), 133–148.

Nagpal, A., Jain, M., & Jain, A., (2019). Unearthing digital financial services and financial inclusion: Empirical evidence from India. *Int. J. Electron Finance, 9*(4), 287–309.

Ni, T. W., (2020). Factors influencing behavioral intention towards adoption of digital banking services in Malaysia. *International Journal of Asian Social Science, 10*(8), 450–457.

Rastogi, S., & Ragabiruntha, E., (2018). Financial inclusion and socioeconomic development: Gaps and solution. *International Journal of Social Economics, 45*(7), 1122–1140.

Saniuk, S., Grabowska, S., & Straka, M., (2022). Identification of social and economic expectations: Contextual reasons for the transformation process of Industry 4.0 into the Industry 5.0 concept. *Sustainability, 14*(3), 1391.

Sankar, B. P., Dash, P., & Leepsa, N. M., (2018). Is demonetization a bridge for digital financial inclusion? *IIMS Journal of Management Science, 9*(2), 93–99.

Siddik, M., Alam, N., & Kabiraj, S., (2020). Digital finance for financial inclusion and inclusive growth. In: *Digital Transformation in Business and Society* (pp. 155–168). Palgrave Macmillan, Cham.

Singh, N. K., Sahu, G. P., Rana, N. P., Patil, P. P., & Gupta, B., (2018). Critical success factors of the digital payment infrastructure for developing economies. In: *International Working Conference on Transfer and Diffusion of IT* (pp. 113–125). Springer, Cham.

Tiwari, T., Srivastava, A., & Kumar, S., (2019). Adoption of digital payment methods in India. *International Journal of Electronic Finance, 9*(3), 217–229.

Vyas, V., & Jain, P., (2021). Role of digital economy and technology adoption for financial inclusion in India. *Indian Growth and Development Review, 14*(3), 302–324.

Wibella, N., Fahmi, I., & Saptono, I. T., (2018). Factors affecting consumer acceptance of digital financial inclusion; anecdotal evidence from Bogor City. *Independent Journal of Management & Production, 9*(4), 1338–1353.

Xu, X., Lu, Y., Vogel-Heuser, B., & Wang, L., (2021). Industry 4.0 and Industry 5.0—Inception, conception, and perception. *Journal of Manufacturing Systems, 61*, 530–535.

CHAPTER 9

An Empirical Study on Accounting Conservatism, Integrated Reporting in India, and Corporate Innovation: Evidence from Industry 5.0

FARIDA RASIWALA[1] and INDRAJIT GHOSAL[2]

[1]Fourtune Institute of International Business, New Delhi, India

[2]Associate Professor, School of Management and Commerce, Brainware University, Barasat, Kolkata, West Bengal, India

ABSTRACT

The purpose of this study is to identify the factors that can influence the adoption of integrated reporting (IR) in emerging markets.

The study is based on a survey of 43 companies, excluding financial and banking companies, which consist of small, medium, and large-scale companies. A questionnaire has been prepared, and responses have been collected from companies that have adopted IR. Factor analysis has been applied to understand the factors impacting the adoption.

The findings of the study show that large-scale companies are adopting IR rapidly compared to medium-scale or small-scale companies. The top two factors identified as the most important factors influencing the adoption of IR are "increased employee engagement" and "concise form of reporting" among other factors.

This research has been conducted with the aim of identifying the main factors that will influence the early adopters of IR in India. This chapter

Sustainable Strategic Business Infrastructure Development and Contemporary Digital Practices in Industry 5.0. Indrajit Ghosal, Bikram Prasad, Nilanjan Ray, & Lalit Kumar Joshi (Eds.)

will serve as a revelation for all other companies that have not yet adopted IR, as they will come to understand the benefits of adopting IR.

This study shed light on prominent factors which can boost adoption of IR in an emerging market context. This provides empirical evidence on the current status of reporting in emerging market such as India and also gathers possible explanations for pressure and challenges faced by Indian companies in the era of high stakeholder expectations.

9.1 INTRODUCTION

Today, the world is witnessing a change in the way business operates and how value is created. With globalization, financial crises, the impact of a global pandemic, the need for more transparent reporting, reduced resources, and increasing environmental concerns, integrated reporting (IR) is a logical solution to many corporate reporting issues. The International Integrated Reporting Council (IIRC), London[1], which aims to present a concise, accurate, and user-friendly annual report developed IR. This reporting tool represents an evolution of environmental and sustainability reports, combining financial and non-financial disclosures (Eccles & Krzus, 2010). It aims to address the changes expected in corporate reporting and adapt to the interconnectedness of today's globalized world (Adams, 2015a). Following the financial crisis of 2008, globalization, and new regulations, corporate reporting was expected to provide information required by various stakeholders, adhering to a range of national and international rules and guidelines (Miller & Skinner, 2015; von Carl-Christian Freidank et al., n.d.). This trend is expected to continue to rise in the post-pandemic era (COVID-19). The growing demand for information disclosures has led to an increase in both voluntary and mandatory reports in emerging economies like India. The use of voluntary disclosure is on the rise to ensure accountability, transparency, and access to more information (Busco et al., 2013; Dumay & Rooney, 2016). In India, this has resulted in an expansion of the number of pages in annual reports. However, the main weakness of corporate reporting lies in the complexity caused by the multitude of different reports presented (von Carl-Christian Freidank et al., n.d.). Furthermore, the increasing number of documents complicates separate publication and makes it challenging for users to understand and extract relevant information.

[1] https://integratedreporting.org/ (accessed on 15 January 2024).

Two key organizations formed the IIRC–the prince's accounting for sustainability[2] and Global Reporting Initiative (GRI)[3] in 2010 to overcome the shortcomings of current corporate reporting practices. IIRC was formed with the aim of developing an internationally recognized approach to IR (International Integrated Reporting Committee, 2011). In 2004, the Prince's Accounting for Sustainability Project (A4S) was initiated with IR.

IR builds on existing corporate reporting practices and proposes to bring together material information about an organization's strategy, governance, performance, and prospects in a way that reflects the commercial, social, and environmental context within which it operates (International Integrated Reporting Committee, 2011, p. 2).

The framework aims to present the firm's outcomes in the short, medium, and long-term. The concept of integrated thinking is also derived from IR, which suggests that a managerial revolution within the company is necessary for the implementation of IR (Idowu, 2019; Stubbs & Higgins, 2014; Thomson, 2015).

Precisely, Integrated Reporting (IR) is guided by the approach of integrated thinking. IR also serves as a catalyst for sustainable business practices (Eccles & Krzus, 2010). Several studies indicate that IR incorporates sustainability practices into management strategies and contributes to addressing environmental, social, and governance (ESG) issues (Black Sun, 2012).The best part of IR is that it interacts between this different information and allows the creation of value over time (de Villiers et al., 2014; Ibrahim Mahmood Al-Nuaimi et al., 2017). Its emphasis on different aspects of the company's performance provides a complete view of value creation (Cheng et al., 2014).

Research is needed to understand various facets of Integrated Reporting (IR), especially in emerging economies. In India, the adoption of IR began with Kirloskar Brothers Ltd, who were the first to adopt IR in 2013–2014, followed by Tata Steel in 2014–2015. On 6th February 2017, the voluntary adoption of the <IR> framework was notified by the Indian Capital market regulator. Since then, companies in India have started adopting IR. Eccles & Serafeim (2017) have pointed out the lack of connectivity between strategies, business models, and outlook as a critical point that

---

[2]It was established by The Prince of Wales in 2004 for making sustainable business decision for sustainable growth. https://www.accountingforsustainability.org/(accessed on 15 January 2024).
[3]GRI was formed in 1997 for supporting and promoting sustainability reporting. https://www.globalreporting.org/(accessed on 15 January 2024).

hinders the company's adoption of IR. It has also been stated that while many companies have started adopting IR, voluntary disclosures of ESG reporting combined with mandatory financial reporting still take precedence (Dumay, 2016a; Dumay et al., 2016).

In the current context, where the company's leadership aims to integrate profit earning and sustainability issues, Integrated Reporting (IR) appears to be the most suitable approach for corporate reporting in India. IR primarily focuses on stakeholders, particularly investors, as Indian companies seek to embrace IR for improved financing options. Transparency in the annual report enables investors and managers to make well-informed decisions (Burke & Clark, 2016; Gasperini, 2014). This study aims to address the gap by identifying the factors that have influenced Indian companies to adopt IR. Considering that the International Integrated Reporting Council (IIRC) is currently updating its framework, with revisions expected by December 2020, this chapter is timely in terms of its placement.

Secondly, there has always been a debate among academics and researchers about IR and how IIRC has evolved. There have been studies that support IR (Adams, 2015a; Cortesi & Vena, 2019a; Melloni et al., 2017) and a few that have a critical opinion about IIRC (Flower, 2014, 2015a; La Torre et al., 2020; Thomson, 2015). In a very recent study by de Villiers & Sharma (2020), it has been argued that the possibility of IR replacing traditional reporting is very low (de Villiers & Sharma, 2020).

Amidst these mixed responses in various studies, it is very difficult to identify the factors that led to the adoption of IR in a country like India. Since the adoption of IR is voluntary, it is crucial to understand the factors that led to its adoption in India. Do companies see any benefits in adopting IR?

Based on the literature review and the existing level of knowledge in this regard, this study addresses the research question of what factors have contributed to the adoption of IR in an emerging economy like India.

This study will be useful for organizations that wish to learn from early adopters about the motivations and challenges of transitioning to an IR framework.

9.2 LITERATURE REVIEW AND RESEARCH GAP

It is well-documented that today's annual report and the value of the firm exist in intangible assets, making their reporting crucial (Dumay, 2016a;

Dumay & Rooney, 2016; Mouritsen et al., 2001). Dumay (2016b) also emphasizes that intellectual capital (IC) has been losing its significance, and therefore, Investor Relations (IR) can learn from IC and capture the company's attention. He compares IR with the closely related field of IC, stating that if we remove manufacturing, financial, and natural capital from the IR framework, the remaining three are the same as IC's three capital.

With the passage of time, the framework also met with criticism (Flower, 2015a; Perego et al., 2016; Thomson, 2015). Some research has also shown that IR is a unilateral, narrow approach for evaluating and reporting sustainability concerns within a firm (Brown & Dillard, 2014).

Nonetheless, many researchers have focused on the theoretical analysis of the framework (Adams, 2017; Haller & van Staden, 2014), while few have explored its relation with other frameworks (Dumay, 2016a; Maas et al., 2016), its merits in relation to sustainable development goals (SDGs) (Adams, 2017; Adams & Simnett, 2011), and its connection with Corporate Social Responsibility (CSR) and business outcomes (Jensen & Berg, 2012; Soyka, 2013). Integrated Reporting (IR) has been proposed as a solution to combine financial reporting, sustainability reporting, CSR activities, and business outcomes. IR is considered a forward-looking reporting approach compared to traditional financial reporting, which tends to be more backward-looking (Kılıç & Kuzey, 2018). IR is seen as a combination of current accounting practices (Velte & Stawinoga, 2017; Veltri & Silvestri, 2015), including public accounting, accounting for sustainability, corporate governance, and conventional reporting.

In India, Investor Relations (IR) is generally considered as voluntary disclosure. Voluntary disclosure is not an obligation, but rather a means to enhance transparency and accountability for stakeholders (Coluccia et al., 2016; Ho & Wong, 2003). IR has been included in the literature on voluntary disclosure due to studies conducted by Abeysekera (2013), Lee & Yeo (2016), and Steyn (2014). These studies have demonstrated the various advantages of IR, such as promoting comprehensive reporting that informs stakeholders about an organization's involvement in political, social, environmental, and governance-related activities. Furthermore, IR provides a holistic view of a business's ability to create value and ensure sustainability. The 2014 EY survey identified several benefits of IR, including improved capital allocation by stakeholders, long-term performance metrics, and insights into the organization's role in society and the effective utilization of funds. Additionally, IR enables businesses to enhance the transparency of both internal and external information, refine policy definition, and optimize reporting. Krzus (2011) identified four key implications of IR that enhance modern reporting, including clearer outcomes, more effective decision-making, increased engagement, and reduced risk of losing credibility.

The existing literature (Wild & van Staden, 2013) reveals that Integrated Reporting (IR) is primarily adopted by large public companies, particularly within the financial services sector, and does not encompass industries with significant social and environmental impacts. Other related studies have indicated that the financial crisis of 2008–2009 was a key driver for the increased demand for non-financial reporting. The theoretical framework for IR is considered a means of aligning with the current reporting approach to achieve an integrated global framework, encompassing interactions with stakeholders and the financial, social, environmental, and governance implications (Abeysekera, 2013). Additionally, research (Durak, 2013) has demonstrated that businesses are more likely to embrace Integrated Reporting (IR) in nations with high societal standards, an abundance of information, and strong governance frameworks. Other difficulties with IR have been noted, such as a lack of consensus on standards and assurance methodologies, a poor understanding of the connection between financial and non-financial results, and a lack of regulations that are appropriate for an effective presentation of IR (Krzus, 2011). Many researchers have tried to determine the factors influencing IR adoption (Galani et al., 2012). Frias-Aceituno et al. (2014) stated that one of the key factors in the creation and distribution of IR is the size of the firm. Frías-Aceituno et al. (2013) have linked IR adoption with the legal status of the country. The benefits and motives of reporting have been analyzed (Farneti & Gurthrei, 2009) through a survey using a semi-structured interview with the Australian public sector, which is known for its "good reporting practices." The findings of a few studies concluded that further clarification of the methodology and future guidance is intended to enhance the application and accuracy of these reports (Dumay et al., 2010).

With various studies focusing on the advantages and challenges of Investor Relations (IR), we observe an increasing adoption of IR worldwide, including in India. Whether this adoption is driven by IR being a rational choice for reporting in the face of current challenges (Adams, 2015b; Churet & Eccles, 2014; Serafeim, 2015), or as a response to pressure from investors and the market (Stubbs & Higgins, 2014, 2018; van Bommel, 2014), or simply as a result of current trends (Flower, 2015b), we can see a rising trend in adoption. Therefore, this chapter aims to identify the factors that are contributing to this increase in adoption in India.

In India, there is a strong emphasis on transparent and concise reporting. In February 2017, SEBI encouraged companies to voluntarily adopt Integrated Reporting (IR) (SEBI, 2017). Sustainability reporting has faced criticism for its lack of social, environmental, and governance information (de Villiers et al., 2014; Solomon & Solomon, 2006). The interconnection between different frameworks and reports is crucial. This interconnection would enhance relevance and facilitate decision-making (Hopwood et al., 2010). Robertson & Samy (2015) conducted a study on senior managers of firms in the UK and found that sustainability reporting lacks linkage and does not adequately support decision-making.

9.3 RESEARCH METHODOLOGY

To address the research question and based on the literature review, a questionnaire-based survey was conducted to understand the factors that led to the adoption of Investor Relations (IR). Companies that have adopted IR were selected for this survey-based study. After analyzing the annual reports of the Top 500 companies (ET 500 companies–2020), it was found that 67 companies are following IR. Since our study focuses on identifying factors for adopting IR, a questionnaire was circulated to these 67 companies. The questionnaire was sent via mail, followed by a follow-up mail and telephonic calls. A total of 43 filled questionnaires were received, which form the sample for this study.

9.3.1 DESIGN, PILOT TESTING, AND RELIABILITY TESTING OF THE QUESTIONNAIRE

The questionnaire was designed based on a literature review (Ghani et al., 2018; Nduneseokwu et al., 2017; Zhou et al., 2017). The questionnaire consists of three sections. Each question is framed using a 5-Likert scale, with 1 indicating "strongly agree" and 5 indicating "strongly disagree." The questions focus on the reasons for implementing IR. Section A covers the demographic information of the respondents, while Section B covers the benefits of adopting IR. Section C addresses various risk and assurance-related issues that could promote the implementation of IR.

The reliability of the study is tested using Cronbach's Alpha. The KMO test is used to determine whether the data can be used for factor analysis.

9.4 RESULTS AND ANALYSIS

Firstly, the results related to the first objectives are discussed. The number of respondents who participated in the study was 43, belonging to 43 companies in India. The collected data was significant because the total number of companies following IR in India was found to be 67. The results are divided into various subsections, which include response rate, descriptive statistics, reliability analysis, and factor analysis (Table 9.1).

TABLE 9.1 Response Rate

Category	Questionnaire Sent	Questionnaire Response	Response Rate
Respondent wise	67	43	64.17%
Company wise	67	43	64.17%

Table 9.2 presents a summary of the demographics of the respondents. Only one questionnaire was sent to each company, addressed to either the CEO (Chief Executive Officer) or CSO (Chief Sustainability Officer). However, it was observed that they forwarded it to their subordinates, who then filled out the questionnaire.

Experience is key to comprehending any subject and responding appropriately. According to the survey, 95.34% of respondents in the sector have more than 15 years of experience. Similar results were found in surveys conducted in Italy and Malaysia on IR, where 88.1% and 96% of respondents, respectively, have worked for their employers for more than five years (Federica et al., 2016; Joshi et al., 2019). Analyzing the industries that responded to the questionnaire, it can be concluded that Metals and Mining had the highest response rate, followed by Consumer Durables, Gems and Jewelry, Chemicals, Pharmaceuticals, Telecommunication, and Tourism and Hospitality. Table 9.2 clearly shows that large-scale companies are pioneers in adopting IR, accounting for 81.39% of the respondents, followed by medium-scale companies. There were no small-sized companies included in the study. The companies were categorized as large, medium, and small-scale based on market capitalization.

TABLE 9.2 Demographic Profiles of the Respondents

	F	Percentage (%)
	(i) Type of Firm	
Metals and mining	8	18.6%
Automotive	6	13.9%
Consumer goods	6	13.9%
Oil and gas	6	13.9%
Utilities	7	16.2%
Others	10	23.2%
	(ii) Experience of Respondent	
Below 15 years	2	4.65%
Above 15 years	41	95.34%
	(iii) Size of Firm	
Large size	35	81.39%
Mid-size	8	18.60%
Small size	0	0%

After conducting the demographic study, a reliability test was applied to the data. Cronbach's alpha was used to examine the internal consistency (Table 9.3).

TABLE 9.3 Cronbach's Alpha

Average interitem covariance	0.0270333
Number of items in the scale	16
Scale reliability coefficient	0.6318

The result of Cronbach's Alpha was 0.6318. Since Cronbach's alpha is greater than 0.5, it indicates the best results for the factorability of the items. According to Goforth (2015), coefficients above 0.6 are always considered acceptable (Table 9.4).

TABLE 9.4 Kaiser-Meyer-Olkin Measure of Sampling Adequacy

Variable	KMO
The numbero ~ n	0.4250
There is mor ~ o	0.5474
Integrate ~ as	0.4387
Therewillb ~ p	0.3494
Therelevan ~ o	0.3380
Therewillb ~ e	0.3833
Thereisreq ~ n	0.5130
Integrated ~ e	0.6886
Theregulat ~ m	0.4746
Thepresent ~ d	0.6621
Integrated ~ r	0.3586
Integrated ~ i	0.3627
Integrate ~ es	0.3543
Integrated ~ o	0.5482
Doyouthink ~ n	0.3462
Overall	0.4438

The Kaiser-Meyer-Olkin (KMO) measure of sampling adequacy indicates that the overall sampling adequacy is above 0.4 for all cases. Therefore, the sample size is sufficient for conducting factor analysis.

Table 9.5 shows the eigenvalue and variance of various factors in the study. According to the Kaiser extraction rule, factors with eigenvalues greater than 1 should be extracted (Boyd, 2013).

The extraction step of factor analysis involves determining the minimum number of shared factors that would adequately account for the correlations between the observed variables. The results of the factor analysis are presented in Table 9.5, and the eigenvalues are visually represented in Figure 9.1. Six factors have been extracted using the eigenvalues.

The factors and their loadings are justified after determining the number of factors to be retrieved. The data rotation technique is applied for this purpose. Factor rotation is mostly used to decompose data and improve variable understanding. The Varimax rotation approach produces a more accurate and simpler-to-understand outcome. The Varimax Rotation method is applied in this research.

TABLE 9.5 Eigen Values and Cumulative Percentage

Statement. No.	Attributes	Initial Eigen Values	Percentage of Variance	Cumulative (%)
S1	Reduced financial and non-financial reports.	3.009	18.808	18.808
S2	Increased transparency in corporate reporting.	2.429	15.181	33.989
S3	Improved number of information.	2.041	12.757	46.747
S4	Upsurge in company's reputation.	1.626	10.163	56.910
S5	Proliferation in employee's engagement.	1.488	9.299	66.209
S6	Enhance accounting information for economic decision-making.	1.129	7.057	73.266
S7	Augmented the accessibility of funds to Indian companies.	0.922	5.763	79.029
S8	Need additional resources to implement integrated reporting.	0.685	4.280	83.309
S9	Integrated reporting rise the gross margin of companies.	0.629	3.930	87.239
S10	Regulatory bodies need to introduce new standard for audit.	0.497	3.105	90.344
S11	Company leaders required skill.	0.407	2.546	92.890
S12	Improves organization's focus on integrating risks.	0.326	2.040	94.930
S13	Improving risk assessment.	0.293	1.830	96.760
S14	Encourages managers to inter-relate strategy, risks, and opportunities of the company.	0.268	1.676	98.435
S15	Brings culture of inserting risk management into the strategic planning process.	0.165	1.030	99.465
S16	Finance and the accounting personnel requisite knowledge of IR framework.	0.086	0.535	100.000

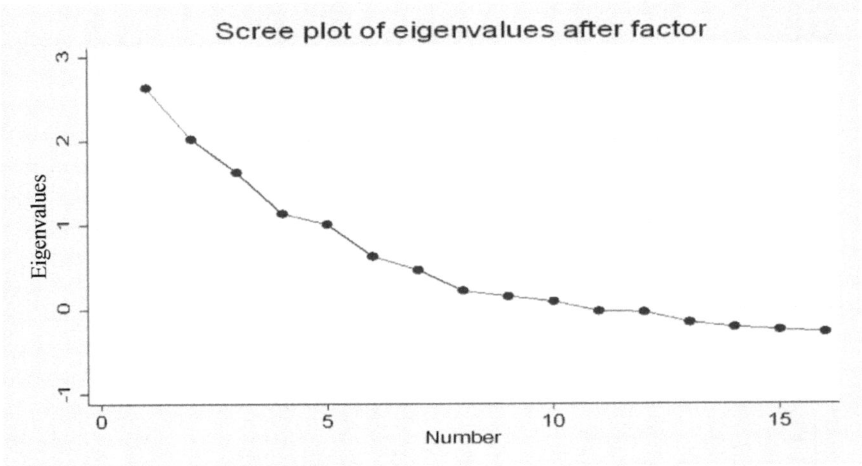

FIGURE 9.1 Scree plot of eigen value.

Factor loadings for the three components that make up factor 1 range from 0.585 to 0.740. The components of factor 1 are S1, S3, and S11. These factors account for 34.65% of the total variation. The eigenvalues of the second, third, fourth, fifth, sixth, and seventh factors are 2.429014, 2.041185, 1.626098, 1.487874, 1.129103, and 1.129103, respectively. The total variation explained by all six components is 100%, which is a significant amount and validates the study.

Using factor analysis, six new factors are effectively created and assigned as variables influencing the uptake of IR in India. The new names for these factors and their constituent parts are shown in Table 9.6. When the variance is extracted, the first factor exhibits the highest percentage of variance, and the subsequent factors have also been sorted in a similar order.

The factors are named after grouping the key variables in explaining a typical attribute IR. The grouping of variables with a coefficient greater than 0.5 is shown in Table 9.6.

Table 9.7 depicts that the first factor identified with reduced financial and non-financial reports, improved number of information, and company leaders requiring skill to present the new information. These attributes can be grouped under factor-I and termed as "Concise reporting." These factors indicate that IR is adopted because of the concise form of reporting (Melloni et al., 2017). In India, there are many financial and non-financial

TABLE 9.6 Rotated Factor Analysis

Statement. No.	Attributes	Factor Loadings					
		F1 Concise Reporting	F2 Informed Decision Making	F3 Increased Transparency	F4 Increased Company's Reputation	F5 Increased Employee's Engagement	
S1	Reduced financial and non-financial reports	0.740	–	–	–	–	–
S3	Improved number of information	0.586	–	–	–	–	–
S11	Company leaders required skill	0.697	–	–	–	–	–
S6	Enhance accounting information for economic decision-making	–	0.530	–	–	–	–
S12	Improves organization's focus on integrating risks	–	0.521	–	–	–	–
S14	Encourages managers to inter-relate strategy, risks, and opportunities of the company	–	0.708	–	–	–	–
S9	Integrated reporting rise the gross margin of companies	–	0.555	–	–	–	–
S2	Increased transparency in corporate reporting	–	–	0.502	–	–	–
S13	Improving risk assessment	–	–	0.837	–	–	–
S4	Upsurge in company's reputation	–	–	–	0.702	–	–
S15	Brings country of inserting risk management into the strategic planning process	–	–	–	0.670	–	–
S5	Proliferation in employee's engagement	–	–	–	–	0.651	–

TABLE 9.6 *(Continued)*

Statement. No.	Attributes	Factor Loadings					
		F1 Concise Reporting	F2 Informed Decision Making	F3 Increased Transparency	F4 Increased Company's Reputation	F5 Increased Employee's Engagement	
S16	Finance and the accounting personnel requisite knowledge of IR framework	–	–	–	–	0.551	–
S10	Regulatory bodies need to introduce new standard for audit	–	–	–	–	0.697	–
	Eigen Value	**5.457**	**2.352**	**2.722**	**1.791**	**1.574**	**0.497**
	Cumulative Variance	**34.111**	**14.703**	**17.011**	**11.193**	**12.939**	**3.105**

TABLE 9.7 Summary of the Factor Categorization

Key Factors of an IR					
F1 Concise Reporting	**F2 Improved Decision-Making Process**	**F3 Increased Transparency**	**F4 Increased Company's Reputation**	**F5 Increase Employee Engagement**	**F6-New Audit Standards**
Reduced financial and non-financial reports.	Enhance accounting information for economic decision-making.	Increased transparency in corporate reporting.	Upsurge in company's reputation.	Proliferation in employee's engagement.	Regulatory bodies need to introduce new standard for audit.
Improved number of information.	Improves organization's focus on integrating risks for better decision making.	Improving risk assessment with more transparent disclosures.	Brings culture of inserting risk management into the strategic planning process.	Finance and the accounting personnel requisite knowledge of IR framework.	–
Company leaders required skill to present new information.	Encourages managers to inter-relate strategy, risks, and opportunities of the company.	–	–	–	–
–	Integrated reporting rise the gross margin of companies	–	–	–	–

disclosures expected from the company. This increases the number of pages of the report. IR is always sought after for its concise reporting structure where all the relevant information is presented without any duplication. The second factor is designated as performance based on the loaded variables. Three variables in this category are important with high factor loading. This indicates that among various feature variables, they enhance accounting information that can help management make strategic decisions, strengthen the organization's attention on risk mitigation, allow executive management to seek inter-related policies, threats, and opportunities, and raise the gross margins of companies. These attributes can be grouped under factor 2 and are named "Improved decision making."

The third factor reveals that Investor Relations (IR) enhances transparency in Corporate Reporting and improves risk evaluation, which is categorized as "Increased Transparency" under factor 3. This factor emphasizes the importance of recognizing transparency as a crucial aspect. On the other hand, the fourth factor highlights the rise in business credibility and the integration of risk management into the strategic planning process. These attributes are grouped under factor 4 and referred to as "Increased Company's Reputation." Therefore, business goodwill is identified as a significant factor for adopting IR in India.

The fifth factor acknowledges the proliferation of employee engagement and the importance of finance personnel with requisite accounting knowledge in the IR Framework. This factor indicates that with increased goodwill, the company will be able to attract more good employees, leading to improved employee engagement. These factors have been grouped under factor 5 and are named "Increased employee engagement."

The sixth factor outlines that regulatory bodies need to introduce a new standard for audits. This factor highlights the need for new audit regulations to increase audit assurance in integrated reporting (IR). The International Integrated Reporting Council (IIRC), in its framework, has provided broad guidelines for IR. Therefore, companies may adopt IR in their own way, which may present some challenges during the audit of their annual reports. Table 9.8 reflects the ranking given to various factors identified by the factor analysis.

From Table 9.8, it can be concluded that increased employee engagement has the highest score, followed by concise reporting. This indicates that in India, increased employee engagement is the most important factor for adopting IR. After this, concise reporting is the second-ranked factor. Increased transparency, increased company's reputation, improved decision-making process, and new audit standards are the other factors arranged according to their rank.

TABLE 9.8 Ranking of Factors

Factor	Rank
Concise reporting	2
Improved decision making	5
Increased transparency	3
Increased company's reputation	4
Increase employee engagement	1
New audit standards	6

9.5 CONCLUSION, RECOMMENDATIONS, AND LIMITATION OF THE STUDY

This research has been conducted with the aim of identifying the main factors that will influence the adoption of Investor Relations (IR) in emerging economies like India. The companies that have adopted IR in India are divided into three categories: large, medium, and small-sized companies for this chapter.

The results of this study suggest that the adoption rate of IR is high in large-sized companies. This indicates that large companies are more inclined to adopt IR compared to small and medium-sized enterprises. IR is an effective tool for enhancing understanding of the relationship between financial and non-financial variables that influence business performance, as well as how a business can generate sustainable value in the long term. In the era of the global COVID-19 pandemic, IR offers more concise and transparent reporting compared to conventional approaches. With globalization, financial crises, and the impact of a global pandemic, IR has become a necessity for reporting with reduced resources and increasing environmental concerns.

Thus, this chapter aims to identify the factors or advantages that influence a company's decision to adopt an integrated report. Several factors were derived from the survey, but only the following factors were found to have the highest influence based on factor loading: increased employee engagement and concise reporting. Other factors that affect the company's decision were ranked as follows: increased transparency, improved reputation of the company, enhanced decision-making processes, and new audit standards.

This chapter also focuses on the difference in cultural backgrounds between developed and emerging economies. This was also analyzed by Cable & Patel (2000), who suggested that the cultural context stimulates the adoption of new accounting and reporting practices. This concept was also studied by Gray & Vint (1995). The adoption of Integrated Reporting (IR) in an emerging economy like India is crucial due to the number of potential companies that may adopt it. Additionally, it provides the necessary assurance to global investors to invest in growing economies.

The result of the study has implications for adaptors, preparers, academics, and regulators. For academics, this chapter contributes to the existing literature by providing evidence on the current status of Integrated Reporting (IR) in the emerging market. For adaptors and preparers of integrated reports, this study clearly identifies the benefits and challenges of adopting the IR framework. In India, regulators are placing significant emphasis on developing new regulations for non-financial reporting. SEBI is introducing a new format called BRSR (Business Responsibility Sustainability Reporting) in 2021. This study provides valuable insights to regulators, helping them formulate more relevant regulations.

There are certain limitations to the study as it only provides a snapshot of Investor Relations (IR) in India. IR is still in its early stages of adoption in India, resulting in a smaller number of companies that have adopted it, making our sample size relatively small. The low adoption rate also reduces the validity of the data. Further studies can focus on the drivers of IR adoption and explore the status of IR adoption in other emerging economy countries.

KEYWORDS

- **concise reporting**
- **employee engagement**
- **Industry 5.0**
- **integrated reporting**
- **JEL classification**

- **large-scale company**
- **M14**
- **M41**
- **pilot testing**
- **transparency**

REFERENCES

Abeysekera, I., (2013). A template for integrated reporting. *Journal of Intellectual Capital.* https://doi.org/10.1108/14691931311323869.

Adams, C. A., & Frost, G. R., (2008). Integrating sustainability reporting into management practices. *Accounting Forum.* https://doi.org/10.1016/j.accfor.2008.05.002.

Adams, C. A., (2015b). The international integrated reporting council: A call to action. *Critical Perspectives on Accounting.* https://doi.org/10.1016/j.cpa.2014.07.001.

Adams, C. A., (2017). Conceptualizing the contemporary corporate value creation process. *Accounting, Auditing and Accountability Journal.* https://doi.org/10.1108/AAAJ-04-2016-2529.

Adams, S., & Simnett, R., (2011). Integrated reporting: An opportunity for Australia's not-for-profit sector. *Australian Accounting Review.* https://doi.org/10.1111/j.1835-2561.2011.00143.x.

Boyd, K. C., (2013). Factor analysis. In: *The Routledge Handbook of Research Methods in the Study of Religion.* https://doi.org/10.4324/9780203154281-22.

Brown, J., & Dillard, J., (2014). Integrated reporting: On the need for broadening out and opening up. *Accounting, Auditing and Accountability Journal.* https://doi.org/10.1108/AAAJ-04-2013-1313.

Burke, J. J., & Clark, C. E., (2016). ScienceDirect the business case for integrated reporting: Insights from leading practitioners, regulators, and academics. *Business Horizons.* https://doi.org/10.1016/j.bushor.2016.01.001.

Busco, C., Frigo, M. L., Quattrone, P., & Riccaboni, A., (2013). Redefining corporate accountability through integrated reporting. *Strategic Finance.*

Cable, D., & Patel, C., (2000). Personality and cultural influences on aggressive financial reporting practices. In: *Asian Review of Accounting.* https://doi.org/10.1108/eb060729.

Churet, C., & Eccles, R. G., (2014). *Integrated Reporting, Quality of Management, and Financial Performance.* In SSRN. https://doi.org/10.1111/jacf.12054.

Coluccia, D., Fontana, S., & Solimene, S., (2016). Disclosure of corporate social responsibility: A comparison between traditional and digital reporting. An empirical analysis on Italian listed companies. *International Journal of Managerial and Financial Accounting.* https://doi.org/10.1504/IJMFA.2016.10001410.

De Villiers, C., & Sharma, U., (2020). A critical reflection on the future of financial, intellectual capital, sustainability, and integrated reporting. *Critical Perspectives on Accounting*. https://doi.org/10.1016/j.cpa.2017.05.003.

De Villiers, C., Rinaldi, L., & Unerman, J., (2014). Integrated reporting: Insights, gaps, and an agenda for future research. *Accounting, Auditing and Accountability Journal*. https://doi.org/10.1108/AAAJ-06-2014-1736.

Durak, M. G., (2013). Factors affecting the companies' preferences on integrated reporting. In: *International Journal of Contemporary Economics and Administrative Sciences*.

Eccles, R. G., & Krzus, M. P., (2010). One report – integrated reporting for a sustainable strategy. *Financial Executive*.

Eccles, R. G., & Serafeim, G., (2017). Corporate and integrated reporting: A functional perspective. In: *Corporate Stewardship: Achieving Sustainable Effectiveness*. https://doi.org/10.9774/gleaf.9781783532605_10.

Elkington, J., (1997). *Cannibals with Forks: The Triple Bottom Line of 21st Century Business*. Capstone, Oxford. Capestone, London. https://doi.org/10.4324/9781849773348.

Federica, D., Andrea, G., & Pasquale, P., (2016). Early adopters of integrated reporting: The case of the mining industry in South Africa. *African Journal of Business Management*. https://doi.org/10.5897/ajbm2016.8029.

Frias-Aceituno, J. V., Rodríguez-Ariza, L., & Garcia-Sánchez, I. M., (2014). Explanatory factors of integrated sustainability and financial reporting. *Business Strategy and the Environment*. https://doi.org/10.1002/bse.1765.

Frías-Aceituno, J. V., Rodríguez-Ariza, L., & García-Sánchez, I. M., (2013). Is integrated reporting determined by a country's legal system? An exploratory study. *Journal of Cleaner Production*. https://doi.org/10.1016/j.jclepro.2012.12.006.

Gasperini, F. D. A., (2014). *The Internal Integrated Reporting and The Value Creation: A Case Study Approach*. IKFAD.

Ghani, E. K., Jamal, J., Puspitasari, E., &Gunardi, A., (2018). Factors influencing integrated reporting practices among Malaysian public listed real property companies: A sustainable development effort. *International Journal of Managerial and Financial Accounting, 10*(2), 144–162. https://doi.org/10.1504/IJMFA.2018.091662.

Goforth, C., (2015). *Using and Interpreting Cronbach's Alpha*. University of Virginia Library.

Havlová, K., (2015). What integrated reporting changed: The case study of early adopters. *Procedia Economics and Finance*. https://doi.org/10.1016/s2212-5671(15)01624-x.

Ho, S. S. M., & Wong, K. S., (2003). Preparers' perceptions of corporate reporting and disclosures. *International Journal of Disclosure and Governance*. https://doi.org/10.1057/palgrave.jdg.2040014.

Hopwood, A., Unerman, J., & Fries, J., (2010). Accounting for sustainability: Practical insights. In: *Accounting for Sustainability: Practical Insights*. https://doi.org/10.4324/9781849776332.

Joshi, P. L., Sallehhuddin, A., Munusamy, P., & Deshmukh, A., (2019). Benefits and adoption of integrated reporting. *International Journal of Knowledge-Based Organizations, 9*(4), 50–64. https://doi.org/10.4018/ijkbo.2019100104.

Keith, T. Z., (2019). Multiple regression and beyond: An introduction to multiple regression and structural equation modeling. In: *Multiple Regression and Beyond: An Introduction to Multiple Regression and Structural Equation Modeling*. https://doi.org/10.4324/9781315162348.

Khemir, S., (2019). Perception of ESG criteria by mainstream investors: Evidence from Tunisia. *International Journal of Emerging Markets*, *14*(5), 752–768. https://doi.org/10.1108/IJOEM-05-2017-0172.

Kılıç, M., & Kuzey, C., (2018). Determinants of forward-looking disclosures in integrated reporting. *Managerial Auditing Journal*. https://doi.org/10.1108/MAJ-12-2016-1498.

Krzus, M. P., (2011). *Integrated Reporting: If Not Now, When?* Blickpunkt: Integrated Reporting.

La Torre, M., Dumay, J., Rea, M. A., & Abhayawansa, S., (2020). A journey towards a safe harbor: The rhetorical process of the International Integrated Reporting Council. *British Accounting Review*, *52*(2), 100836. https://doi.org/10.1016/j.bar.2019.100836.

Mouritsen, J., Larsen, H. T., & Bukh, P. N. D., (2001). Intellectual capital and the "capable firm": Narrating, visualizing, and numbering for managing knowledge. *Accounting, Organizations and Society*. https://doi.org/10.1016/S0361-3682(01)00022-8.

Nduneseokwu, C. K., Qu, Y., & Appolloni, A., (2017). Factors influencing consumers' intentions to participate in a formal e-waste collection system: A case study of Onitsha, Nigeria. *Sustainability (Switzerland)*. https://doi.org/10.3390/su9060881.

Perego, P., Kennedy, S., & Whiteman, G., (2016). A lot of icing but little cake? Taking integrated reporting forward. *Journal of Cleaner Production*. https://doi.org/10.1016/j.jclepro.2016.01.106.

Robertson, F. A., & Samy, M., (2015). Factors affecting the diffusion of integrated reporting – a UK FTSE 100 perspective. *Sustainability Accounting, Management and Policy Journal*. https://doi.org/10.1108/SAMPJ-07-2014-0044.

Stubbs, W., & Higgins, C., (2014). Integrated reporting and internal mechanisms of change. *Accounting, Auditing and Accountability Journal*. https://doi.org/10.1108/AAAJ-03-2013-1279.

Stubbs, W., & Higgins, C., (2018). Stakeholders' perspectives on the role of regulatory reform in integrated reporting. *Journal of Business Ethics*. https://doi.org/10.1007/s10551-015-2954-0.

Thomson, I., (2015). "But does sustainability need capitalism or an integrated report" a commentary on "The International Integrated Reporting Council: A story of failure" by Flower, J. *Critical Perspectives on Accounting*. https://doi.org/10.1016/j.cpa.2014.07.003.

Veltri, S., & Silvestri, A., (2015). The free state university integrated reporting: A critical consideration. *Journal of Intellectual Capital*. https://doi.org/10.1108/JIC-06-2014-0077.

Vesty, G. M., Ren, C., & Ji, S., (2018). Integrated reporting as a test of worth: A conversation with the chairman of an integrated reporting pilot organization. *Accounting, Auditing and Accountability Journal*. https://doi.org/10.1108/AAAJ-08-2016-2684.

Zhou, S., Simnett, R., & Green, W., (2017). Does integrated reporting matter to the capital market? *Abacus*. https://doi.org/10.1111/abac.12104.

CHAPTER 10

Assessment of Investment Priorities by Rural Women

KOPPALA VENUGOPAL

Department of MBA, Aditya Institute of Technology and Management, Tekkali, Srikakulam, Andhra Pradesh, India

ABSTRACT

Investment priorities by rural women are exceptionally significant, as they are generally well-equipped to make savings and profitable choices. Their knowledge of different investment options helps maintain prosperity, risk tolerance, and consistency. This research aims to assess the factors that influence the investment preferences of rural women. The study utilized a mixed method approach, combining qualitative and quantitative methods, with a convenience sample size of 112 participants. The results of the ANOVA analysis revealed the significance and impact of prosperity, risk tolerance, and consistency on investment preferences. Additionally, the study established the strength of each item related to the specified variables, which explain the dependent variable.

10.1 INTRODUCTION

The existence of clear and enabling plans and strategies, the presence of rural authorities, the construction of infrastructure, the transfer of technology, and the availability of agriculture and natural resources such as water, minerals, and tourism resources have created a good opportunity

Sustainable Strategic Business Infrastructure Development and Contemporary Digital Practices in Industry 5.0. Indrajit Ghosal, Bikram Prasad, Nilanjan Ray, & Lalit Kumar Joshi (Eds.)

for the development of rural India. Despite the huge resources available in rural areas and the efforts made by various sectors, the rural community has still not been able to achieve the optimum level of economic growth and attract the required level of domestic and foreign direct investment (FDI) that is commensurate with its economic potential.

The government's focus on policies and practices for the growth of rural areas is evident through recent initiatives and developments such as village electrification, the maiden agriculture export policy (2018), and rural housing. These efforts aim to make the rural community more profitable by increasing income levels. It has been observed that most of the income is saved in traditional saving systems, which provide security but yield lower returns. Rural people tend to prioritize savings through secure banking and postal systems, often hesitating to invest in businesses. This research will select variables from previous research reviews, such as initial capital, amount of borrowed loans, labor and input costs, market problems, skilled labor force, total revenues, number of competitors, and educational level. These variables are expected to have a significant impact on investment behavior, allowing the researcher to identify opportunities and challenges for investment in rural areas.

Emphasizing the role of rural women in investment practices is highly significant, particularly in relation to their thrifty saving attitude compared to the extravagant nature of rural men. Rural women have been found to exhibit a typical attitude towards prosperity, risk tolerance, and consistency when it comes to investing in various available plans. They prioritize saving and reduce expenditure. While several studies have been conducted on rural investments, none have specifically focused on rural women. This gap exists in assessing investment priorities in terms of prosperity, risk tolerance, and consistency.

The novelty lies in the inclusion of women and their investment behavior, which would make a significant contribution to rural investments with small assortments, safeguarding their families from economic emergencies. Currently, most rural women are still limited to traditional investment practices that yield minimal profits and lack transactional-based economic activity. Therefore, the identification and assessment of their behavioral change towards modern practices is a key aspect of this project. When discussing the business practices of rural women, the sustainability of rural micro-level businesses, in particular, poses a major challenge, and family businesses are also not very promising. Small businesses are scarce as most rural areas rely on micro- and tiny family businesses.

Hence, this study carries novelty in terms of relevance and the need to identify the problems and prospects of investment, and to align the investment behavior of rural women with the increased level of benefits and the optimum utilization of government initiatives. Moreover, the objectives of the study align with the thematic area of Psychology and Management, specifically focusing on rural women's behavior in investment.

There have been numerous research studies on investment-related issues, but a significant gap exists in the focus on rural areas and the women community residing there, which are far removed from urbanized areas. Urbanized areas have better access to investment information through various media and sources, enabling them to make informed investment decisions. In contrast, rural investors, especially women, lack awareness of investment opportunities and prospects, face numerous challenges, and often make poor investment choices. Additionally, they are unable to explore the potential benefits of different investment plans. Furthermore, it is crucial to understand the reasons behind the conservative behavior of certain rural women towards modern investment plans and entrepreneurial opportunities.

10.2 OBJECTIVES

The overall objectives in brief and specific of the study are:

- To explore the potential investment opportunities backed by the resources available and utilized in rural India.
- To investigate the magnitude, scope, and depth of challenges faced by rural women in establishing the investment plan.
- To assess the investment behavior of rural women towards the explored opportunities in line with prosperity, risk tolerance, and consistency.

10.3 LITERATURE REVIEW

It has been concluded in a research study that the focus should be on all individual investors and their preferences towards the explored investment product characteristics that suit their financial needs (Rajarajan, 2010). Since importance should be given to every member of the family who has different qualities and ambitions, this may lead to specific and separate choice criteria.

The impact of personal traits has a high significance on the behavior of the investor, which in turn leads to the acquisition of maximum possible knowledge and prospective investment behavior. Additionally, knowledge obtained internally is more imperative than external locus of control. It is also concluded that the level of investment knowledge determines individual investors' decisions (Shanmugham & Ramya, 2012). In general, rural women are considered to have limited exposure and knowledge due to the rural environment. However, the changes brought about by urbanization have been explored to assist rural individuals in selecting advantageous investment options.

Rural people's investment behavior is typically conventional and influenced by their investment culture, which is greatly shaped by the masses. This plays a significant role in the capital formation of the nation as a whole. Any investment is generally planned based on individual characteristics such as attitude, perception, and financial goals (Ramakrishna & Krishnudu, 2009). Despite the exposure to various investment options through traditional and digital media, it is important to study the mindset of rural women, who may exhibit slightly different behavior due to urban transition.

Individuals place a greater emphasis on safety when investing in any financial plan. Therefore, they prefer to invest in national banks and LIC through deposits and policies, which are traditionally considered highly secure. It is also worth noting that investors now evaluate the return and risk associated with their investments (Ramprasath & Karthikeyan, 2013).

Individuals in higher tax brackets tend to opt for small saving plans such as schemes, postal deposits, and secured chits. Rural individuals, in particular, believe that their money is more secure when saved in postal small saving schemes, which have gained trust due to their good interest rates and high assurance (Ganapathi & Anbu, 2008).

Vital conclusions have been drawn regarding the assessment of choice criteria among rural individuals in relation to small saving schemes. These schemes are designed with attributes that cater to rural needs, but the facilities and services provided are inadequate to meet customer convenience. It is suggested that exceptional doorstep services should be offered to attract rural investors (Kasilingam & Jayabal, 2009).

The changing lifestyles of investors, influenced by concurrent knowledge and investment practices, are a crucial factor that significantly impacts investment patterns. Any individual exposed to the investment environment has the opportunity to select a plan based on careful consideration, reliable sources, reference groups, and updated knowledge (Sushant & Bodla, 2009).

The majority of rural individuals highly consider risks and return on investment when making investment decisions. This is often due to conventional thinking and a lack of in-depth knowledge about the best investment options. Other factors such as unwillingness, lack of literacy, and dogmatic behavior also play a role. The theory of planned behavior is used to understand rural investing decisions (Sunil, 2013).

There is a noticeable difference in the perceptions of experts and non-experts regarding risk tolerance and communication. To address this, a program has been developed to re-educate consumers and improve communication. The redefined procedures aim to involve investors more in management, risk assessment, and communication orientation (Kapil, 2012).

According to a study on Perceptions and Perceptual Gaps, most people prefer to invest in bank deposits and postal schemes. However, they strongly disagree with investing in corporate securities, which requires educating investors to develop the right perspective on the advantages of investment schemes (Shollapur & Kuchanur, 2008).

There has been a significant shift in the investment practices of rural people in Kerala state, as they have embraced unconventional investment plans that go against conventional thinking (Sunny & Rajesh, 2014).

Rural women have distinct investment preferences, prioritizing consistency. They are highly skilled in understanding and resolving operational problems related to future financial challenges (Koppala & Saumendra, 2022).

10.4 METHODS

In this study, a descriptive research design and a mixed approach of qualitative and quantitative research were used. The sources of data for this study included both primary and secondary sources. Primary sources involved collecting data from rural women through survey questionnaires. Additionally, the study utilized secondary data sources such as official investment reports, government office reports, magazines, investment guidelines, and other research papers.

The target population for this study consisted of rural women investors in the North coastal districts of A.P. Sample respondents were selected using convenience sampling techniques, as the population of rural women with an inclination towards making investment decisions was undefined. The determined sample size was 112.

To determine the significance of the relationship between the continuous variables considered in the study, an inferential method of analysis was employed. Since the dependent variable is continuous in nature, multiple regression models were utilized using SPSS 24 version software.

10.5 ANALYSIS AND INTERPRETATION

The analysis was conducted to achieve the objectives of the study. The contribution of variables, the impact of factors, and the assessment of items' strength in explaining the dependent variable have been analyzed and interpreted using the following tabulations of model summary, ANOVA, and multiple regression.

As shown in Table 10.1, the R-square values for the variables of Prosperity, Risk Tolerance, and Consistency are 0.002, 0.105, and 0.100, respectively. This indicates that the levels of the items for these three variables contribute 2%, 10.5%, and 10% respectively to the increase in Investment Practices of rural women.

TABLE 10.1 Model Summary

	Model	R	R Square	Adjusted R Square	Std. Error of the Estimate
1	Prosperity	0.046[a]	0.002	–0.025	1.26125
2	Risk tolerance	0.325[a]	0.105	0.064	1.20525
3	Consistency	0.316[a]	0.100	0.075	1.19800

1[a]: Predictors: (Constant), we invest for the future benefits of our family, we opt for all the investment plans for long-term benefits, we reduce our liabilities and invest for long-term benefits.

2[a]: Predictors: (Constant), we don't have much money to take risks, we don't go for risky investments, we feel that investing in business is risky, we feel that investing in land and gold is risky, we feel that any money invested in banks and post office is safe, we feel that anything that gives huge returns is risky.

3[a]: Predictors: (Constant), we invest more in steadily growing investment plans, we invest on a daily, weekly, and monthly basis, continuous investments yield good benefits.

Table 10.2 shows the relationship among the three independent variables: prosperity, risk tolerance, and consistency, and the dependent variable: increase of investment practices of rural women. The F values between the dependent variable and the predictors of prosperity, risk tolerance, and consistency are 0.076, 2.522, and 4.022, respectively. Subsequently, the p values for prosperity, risk tolerance, and consistency are 0.973 (not significant), 0.034 (significant at the 0.05 level), and 0.009 (significant at the 0.05 level), respectively, with the latter being highly significant at the 0.05 and 0.00 levels.

TABLE 10.2 ANOVA[b]

	Model		Sum of Squares	Residual	Mean Square	F	Sig.
1	Prosperity	Regression	0.361	173.391	0.120	0.076	0.973[a]
2	Risk tolerance	Regression	18.320	155.432	3.664	2.522	0.034[a]
3	Consistency	Regression	17.315	156.437	5.772	4.022	0.009[a]

1[a]: Predictors: (Constant), We invest for the future benefits of our family, we choose all the investment plans for long-term benefits, we decrease our liabilities and invest for long-term benefits.

2[a]: Predictors: (Constant), we don't have much money to take risks, we don't go for risky investments, we feel that investing in business is risky, we feel that investing in land and gold is risky, we feel that any money invested in banks and post office is safe, we feel that anything that gives huge returns is risky.

3[a]: Predictors: (Constant), we invest more in steadily growing investment plans, we invest on a daily, weekly, and monthly basis, and continuous investments provide us with good benefits.

[b]Dependent variable is the increase in investment practices of rural women.

Increase in investment practices of rural women (IIPR) = 3.311+ (0.018) P1 + (–0.013) P 2 + (0.029) P3.

An increase in investment practices of rural women, influenced by the variable prosperity, is established in Table 10.3. The items pertaining to Prosperity are as follows: P1 (Investing for the future benefits of the family) is 3.329 (3.311+0.018). In case P1 is better by one unit, the investment practices of rural women will increase by 3.329. Similarly, if the predictors P2 and P3 are better by one unit, the investment practices of rural women increase for P2 (Opting for all the investment plans for long-run benefits) by 3.318, and for P3 (Reducing liabilities and investing for long-run benefits) by 3.360.

The increase in investment practices of rural women is explained by all the items of Prosperity with almost similar weightages. The item "Reducing liabilities and investing for long-run benefits" explains the highest increase with 3.360, followed by "Investing for the future benefits of the family" with 3.329. The least increase is explained by "Opting for all the investment plans for long-run benefits" with 3.318.

TABLE 10.3 Coefficients[a] of Prosperity

Model		Unstandardized Coefficients		Standardized Coefficients	t	Sig.
		B	Std. Error	Beta		
1	(Constant)	3.311	0.501	–	6.612	0.000
	We invest for the future benefits of our family.	0.018	0.057	0.030	0.309	0.758
	We opt all the investment plans for long run benefits.	–0.013	0.117	–0.011	–0.107	0.915
	We reduce our liabilities and invest for long run benefits.	0.029	0.104	0.029	0.282	0.779

[a]Dependent variable: Investment practices of rural women are increased.

The equation for the Increase in Investment Practices of Rural Women (IIPR) is 4.381 + (0.234) RT1 + (–0.264) RT2 + (–0.250) RT3 + (0.006) RT4 + (0.084) RT5 + (0.034) RT6.

An increase in investment practices of rural women, influenced by the variable risk tolerance, is established in Table 10.4. Items pertaining to the variable risk tolerance are as follows: RT1 (Money scarcity to take risks) is 4.615 (4.381 + 0.234). In the case of RT1 being better by one unit, the investment practices of rural women will increase by 4.615. Similarly, if the predictors RT2, RT3, RT4, RT5, and RT6 are better by one unit, the investment practices of rural women increase as follows: RT2 (Unwillingness to risk investments) by 4.117; SMH3 (Investing in Business is risky) by 4.131; SMH4 (Investing in land and gold are risky) by 4.387; SMH5 (Investing in banks and post office is safe) by 4.465; and SMH6 (Anything that gives huge returns is risky) by 4.415.

The increase in investment practices of rural women can be explained by all the items of risk tolerance, with almost similar weightages. The item that explains the highest weightage is "Money scarcity to take risks" with a score of 4.615, followed by "Money invested in banks and post office is safe" with a score of 4.465. The item that explains the least weightage is "Unwillingness to risk investments" with a score of 4.117.

TABLE 10.4 Coefficients[a] of Risk Tolerance

Model	Unstandardized Coefficients		Standardized Coefficients	t	Sig.
	B	Std. Error	Beta		
1 (Constant)	4.381	0.739	–	5.931	0.000
We don't have much money to take risks.	0.234	0.091	0.253	2.577	0.011
We don't go for risk investments.	–0.264	0.121	–0.231	–2.188	0.031
We feel that investing in business is risky.	–0.250	0.124	–0.200	–2.006	0.047
We feel that investing in land and gold are risky.	0.006	0.108	0.006	0.059	0.953
We feel any money invested in banks and post office is safe.	0.084	0.114	0.075	0.743	0.459
We feel that anything gives huge returns is risky.	0.034	0.041	0.084	0.833	0.407

[a]Dependent variable: Investment practices of rural women are increased.

Increase in Investment Practices of rural women (IIPR) = 4.315 + (0.142) C1 + (–0.301) C 2 + (–0.100) C3.

Increase in investment practices of rural women being influenced by the variable prosperity are established in Table 10.5. Items related to Prosperity are as follows: C (Investing more in steadily growing investment plans) is 4.457 (4.315+0.142). If C1 is better by one unit, the Investment Practices of rural women will increase by 4.457. Similarly, if predictors C2 and C3 are better by one unit, the Investment Practices of rural women will increase for C2 (Investing on a daily, weekly, and monthly basis) by 4.041, and for C3 (Continuous investments give us good benefits) by 4.215.

The increase in investment practices of rural women is explained by all the items of Prosperity with almost similar weightages. The item "Investing more on steadily growing investment plans" has the highest explanation with a weightage of 4.457, followed by "Continuous investments give us good benefits" with a weightage of 4.215. The least explanation is given by "Investing on a daily, weekly, and monthly basis" with a weightage of 4.215 (Table 10.5).

TABLE 10.5 Coefficients[a] of Consistency

Model	Unstandardized Coefficients		Standardized Coefficients	t	Sig.
	B	Std. Error	Beta		
1 (Constant)	4.315	0.555	–	7.782	0.000
We invest more on steadily growing investment plans.	0.142	0.100	0.136	1.419	0.159
We invest on daily, weekly, and monthly basis.	–0.301	0.093	–0.308	–3.246	0.002
Continuous investments give us good benefits.	–0.100	0.093	–0.099	–1.077	0.284

[a]Dependent variable: investment practices of rural women are increased.

10.6 RECOMMENDATIONS

1. Some rural women have the intention of reducing unnecessary expenses and saving for future generations. They are interested in investing in prospective investment plans that provide good benefits over a long period of time. Many women prefer not to withdraw from long-term plans offered by banks and post offices, as these plans are aimed at major life events such as their daughters' weddings or building a new house. They connect long-term investment plans with specific goals in their lives. In addition to traditional plans, both governmental and private financial institutions should focus on this group and develop policies that offer long-term benefits, aligning with the needs of rural women.
2. Rural women prefer instant plans such as specific schemes, chits, and short-term savings. This is because most of them earn on a daily basis, and their investments are naturally in small amounts. Investment companies may focus on this group and develop

specific plans for instant collections and short-term investments that align with their wages and the continuity of their work. Policy makers should also consider incorporating certain flexibilities in collecting investments.

3. Most of the rural women who work are poor and receive low wages, which are mostly used for daily expenses. The remaining amount is either invested or saved. Given the challenging circumstances faced by rural women, it is not advisable for them to consider risky investment plans. It is important to refrain from proposing or promoting such plans to this particular group, as they constitute a significant portion of the rural population. It is common sense that when money reserves are limited, the risks taken should also be minimized.
4. In general, rural women feel that investing in business is an option for those who have been in business for a long period of time, and certain communities that are specifically meant for it have all the capabilities to do it and sustain it. Therefore, most women assume that business practices are risky unless one is capable of doing it. Additionally, there have been numerous examples of failed micro-businesses that could not survive due to incompetence. Therefore, the government should educate and train women, especially about the risks and how to overcome conventional thinking about business practices.
5. Rural women assume that investing in gold and land would provide them with consistent and guaranteed profits in the long run. Many women in rural areas purchase gold in the form of jewelry and invest in agricultural fields as land, and they have witnessed the prices of both assets increase exponentially beyond their expectations. However, not all rural women are able to afford such large investments, and they rely on daily, weekly, and monthly investments instead. Once they have accumulated a significant amount of savings, they may consider making long-term investments in land and gold as mentioned above.
6. There has been a keen insight into the steady growth of investment plans among rural women. This is due to their firsthand experience with the negative consequences of poor investments made by their family members. As a result, they are apprehensive about engaging in risky investments and are hesitant to invest in modern

investment plans related to the stock market and contemporary business practices.

KEYWORDS

- **consistency**
- **investment behavior**
- **modern business practices**
- **prosperity**
- **risk tolerance**
- **rural women**

REFERENCES

Ganapathi, R., & Anbu, M. S., (2008). Customer satisfaction of MEPZ–special economic ZONE. *Asia-Pacific Journal of Management Research and Innovation, 4*(1), https://doi.org/10.1177/097324700800400111.

Kapil, S., (2012). Identifying relationship between capital structure and value of the firm for Indian pharmaceutical companies. *The Journal of Contemporary Management Research, 6*(2), 77–85.

Kasilingam, R., & Jayabal, G., (2009). A study on the perception of investors towards small saving schemes. *Paradigm, 13*(2), 1–12. doi: 10.1177/0971890720090202.

Koppala, V., & Saumendra, D., (2022). Assessment of operational functions affecting on the micro and small enterprises' sustainability. *Parikalpana – KIIT Journal of Management, 18*(1), 126–133.

Rajarajan, V., (2010). Indian investors investment characteristics. *Finance India, 24*(4), 1275–1294.

Ramakrishna, R. G., & Krishnudu, C., (2009). Investment behavior of rural investors. *Finance India the Journal of Indian Institute of Finance, XXIII*(4), 1281.

Ramprasath, S., & Karthikeyan, B., (2013). Individual investors' behavior towards selected investments: A study with special reference to Kattumannar. *The International Journal of Business & Management, 1*(6).

Shanmugham, R., & Ramya, K., (2012). Impact of social factors on individual investors' trading behavior. *Procedia Economics and Finance* (pp. 237–246). Published by Elsevier Ltd. Selection and/or peer-review under responsibility of Global Science and Technology Forum.

Shollapur, M. R., & Kuchanur, A. B., (2008). Identifying perceptions and perceptual gaps: A study on individual investors in selected investment avenues. *The ICFAI University Journal of Behavioral Finance*, *5*, 48–64.

Sunil, K., (2013). Investment attitude of rural investors. *Anveshanam: a National Journal of Management*, *1*(1).

Sunny, K. T., & Rajesh, M. N., (2014). *Investment Pattern of Rural Investors in Kerala* (p. 15). Southern Economist.

Sushant, N., & Bodla, B. S., (2009). Impact of investors' lifestyle on their investment pattern: An empirical study. *The ICFAI University Journal of Behavioral Finance, 6*(2), 28–51.

CHAPTER 11

Evaluating the Impact of Physical and Digital Infrastructure for the Development of Tourism Business in North-East India, with Special Reference to Assam

SOUMENDRA NATH BISWAS[1] and PUSPANJALI MOHAPATRA[2]

[1]*Department of Hospitality and Tourism Management, Assam University (A Central University), Silchar, Assam, India*

[2]*Utkal University of Culture, Bhubaneswar, Odisha, India*

ABSTRACT

Attempts have been made in this study to evaluate the impact of physical and digital infrastructure development on the sustainable growth of tourism business in Northeast India, specifically in Assam, and how it influences the socio-economic life of the people in this destination. This chapter begins by understanding the various impacts of physical and ICT-based infrastructure development that affect tourism business in the study area. The key issue of this chapter is to assess the current scenario of physical and digital infrastructure in the state and evaluate its impact on sustainable and eco-friendly tourism development in Assam. The central idea of this chapter is sustainable socio-economic development through information technology. Transport and tourism-related digital infrastructure are major factors shaping development patterns in the Northeastern

Sustainable Strategic Business Infrastructure Development and Contemporary Digital Practices in Industry 5.0. Indrajit Ghosal, Bikram Prasad, Nilanjan Ray, & Lalit Kumar Joshi (Eds.)

zone. Governments at all levels must assume appropriate proactive roles for sustainable development.

11.1 INTRODUCTION

The Northeastern Region of India is one of the most ethnically and linguistically diverse regions in the country. It comprises eight states: Arunachal Pradesh, Assam, Meghalaya, Manipur, Mizoram, Nagaland, Tripura, and Sikkim. This region covers an area of 262,179 square kilometers, which is 7.98% of India's total geographical area, and is divided into 120 districts. The region shares international borders with several neighboring countries, including Nepal (99 km), Bhutan (516 km), China (1,346 km), Myanmar (1,643 km), and Bangladesh (1,879 km). According to the last Census in 2011, the population of the region is approximately 45 million (4.5 crore), accounting for 3.78% of India's total population. The region has a significant concentration of tribal population, with three states having a Scheduled Tribe (ST) population above 85%: Mizoram (94.43%), Nagaland (86.48%), and Meghalaya (86.15%). Each tribe in the region has its own unique traditions of art, culture, dance, music, and lifestyle (Table 11.1).

TABLE 11.1 Population of North-Eastern States of India

State	Male	Female	Total	Percentage Share of India
Assam	1,59,39,443	1,52,66,133	3,12,05,576	2.58%
Tripura	18,74,376	17,99,541	36,73,917	0.30%
Meghalaya	14,91,832	14,75,057	29,66,889	0.25%
Manipur	14,38,586	14,17,208	28,55,794	0.24%
Nagaland	10,24,649	9,53,853	19,78,502	0.16%
Arunachal Pradesh	7,13,912	6,69,815	13,83,727	0.11%
Mizoram	5,55,339	5,41,867	10,97,206	0.09%
Sikkim	3,23,070	2,87,507	6,10,577	0.05%
Northeastern region	2,33,61,207	2,24,10,981	4,57,72,188	3.78%

Source: Population Census (2011).

Due to its remote location, this region is facing a deficiency of infrastructure. This isolated region of India lags behind other regions of the country in many development indicators. Although infrastructure has developed over the years, the region still has a long way to go to meet national standards. Major industries in this region include tea, sericulture, bamboo processing, agriculture, horticulture, rubber, handicrafts, and tourism, among others. This region produces the highest volume of Eri and Muga Silk in the country. In 2019, approximately 100 lakh tourists visited this region, with Assam receiving the highest number of tourists (54.7 lakh), followed by Sikkim (15.6 lakh) and Meghalaya (12.7 lakh). Assam is one of the most fascinating places in northeast India. A major portion of this region remains isolated and unexplored due to its rugged terrain, which includes mountains, hills, rainforests, rivers, and a high density of biodiversity.

The Government of India has initiated the 'Northeast Special Infrastructure Development Scheme' (NESIDS) for the development of infrastructure. This scheme provides financial assistance for water supply, power, transport services, and improving connectivity to boost the tourism sector. Education and health infrastructure development also fall under this sector. The Northeastern Council (NEC) plays a crucial role in developing road and air connectivity in the region. Through the establishment of important regional institutions and various socio-economic infrastructures, NEC is enhancing the quality of life in Northeast India.

The Government has taken important initiatives for the development of the North East region, such as the Special Accelerated Road Development Program for North East (SARDP-NE), projects under the Swadesh Darshan Scheme for Integrated Development of Tourist Circuits and PRASAD Scheme for tourism infrastructure, the National Sports University at Imphal, the development of greenfield airports and the upgradation of airports in different places of North East, and the development of Brahmaputra and 19 new waterways including Barak river (DoNER, 2019). It is crucial to strengthen Information and Communications Technology in this isolated region for the planning, implementation, and monitoring of various development schemes and projects.

11.2 INFRASTRUCTURE DEVELOPMENT (PHYSICAL) IN NORTHEAST

The North-East Special Infrastructure Development Scheme (NESIDS), a centrally funded scheme, has been implemented by the government to develop social and physical infrastructure in the region. In Assam, under the sixth schedule of the constitution, three special development packages have been implemented to address the long-standing demands of the people in these areas. These packages are the Bodoland Territorial Council (BTC), Dima Hasao Autonomous Territorial Council (DHATC), and Karbi Anglong Autonomous Territorial Council (KAATC).

The Northeastern States Roads Investment Program (NESRIP) has taken the initiative to construct/upgrade a total of 433.42 km of roads spread across the 6 North Eastern States of Assam, Manipur, Meghalaya, Mizoram, Sikkim, and Tripura. It was approved in 2011, with a target for project completion by August 2022. With the exception of one road in Tripura, all road projects under NESRIP have been completed.

The Ministry of DoNER launched the Northeast Road Sector Development Scheme (NERSDS) during 2015-2016, with an initial approved budget of Rs. 450.00 crore until March 2017. Subsequently, a proposal for the extension of the scheme was considered, with an expenditure of Rs. 1,000 crore. The scheme has been extended until 2020–2021. Under DoNER, there are two corporations in operation. The first is the Northeastern Handicrafts and Handlooms Development Corporation Ltd. (NEHHDC), which focuses on the development and promotion of handicrafts and handloom products, providing market linkages to artisans and weavers through exhibitions, trade bazaars, and various events. The second is the Northeastern Regional Agricultural Marketing Corporation (NERAMAC), which is involved in food processing and trading of agricultural products. The government has taken various measures to improve air, road, rail, waterways, and telecom connectivity in the Northeast region, resulting in visible growth in different parts of the region. Special schemes for infrastructure development are being implemented by the BTC, Dima Hasao Territorial Council (DHTC), and KAATC under State Plan Assistance.

There are several inter-state roads, mostly constructed by NEC and BRO, in this region. Many of these roads require maintenance to ensure smooth interstate connectivity. Unfortunately, many economically important roads have been neglected for various reasons:

- In some cases, these interstate roads may be vital for the people of one state but neglected by other connected states.
- Some roads are important due to security issues and their strategic locations in the border areas of other countries.
- Few roads are important because of accessibility towards marketplace, cities, economic importance, tourist place, etc.

Infrastructure development, especially the transportation network, is the lifeline for any civilization. There is a need for more investment in the following sectors of this region:

- Transport and communication;
- Infrastructure for digital communication;
- The power sector, especially from alternative natural energy sources;
- Agriculture and allied sector including tea gardens;
- Animal husbandry, including dairy and sericulture;
- Irrigation;
- The tourism sector, especially in ecotourism, rural tourism, agro-tourism, tea tourism, etc.;
- Medical and health sector;
- Science and technology;
- Education sector;
- Medical and health sector.

11.2.1 THE ROLE OF THE NORTHEASTERN COUNCIL (NEC) IN INFRASTRUCTURE DEVELOPMENT.

The NEC was formed as a statutory body by an Act of Parliament for the socio-economic development of the region. In order to actively support this development, various infrastructure-related works such as road and air connectivity, the establishment of significant regional institutions, and the creation of various socio-economic infrastructures have been undertaken in this region. NEC is also actively engaged in capacity building, training, and development, as well as boosting digital infrastructure in the region. The major areas of infrastructure development identified by NEC include transport and communication, industries, tourism, power, agriculture, irrigation, medical and health, science and technology, and HRD and employment, among others.

11.3 DIGITAL INFRASTRUCTURE IN ASSAM

For overall development in the region, along with physical infrastructure, digital infrastructure is also important. Assam Government has initiated digitally empowered social and knowledge hub as per the Digital India mission. It endeavors to digitally connect all the villages of the state and indication of digital revolution in Assam. In entire northeast and especially in interior part of Assam the IT infrastructure is not as per the other areas of the country. The internet bandwidth is improving and the companies like BSNL, Jio, Airtel, and VI are trying to expand their business in this region. The concept of 'Digital club' is also coming here. For infrastructure development in the entire state, digital Assam is working on all the nine pillars of digital India (Figure 11.1).

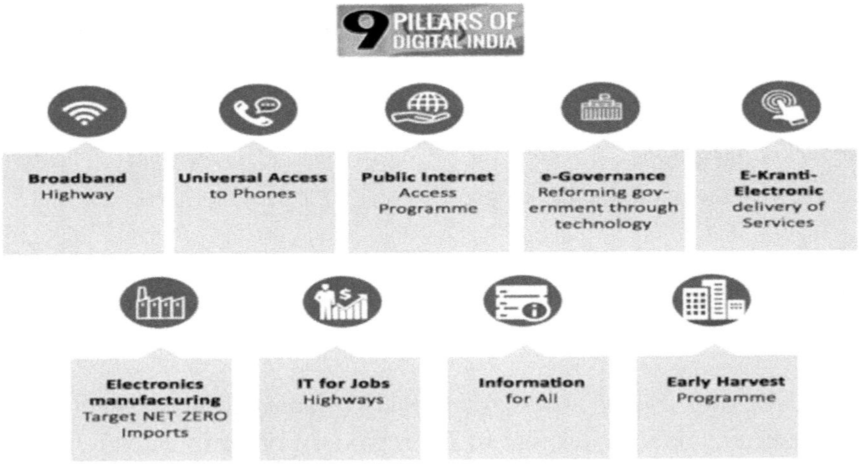

FIGURE 11.1 Nine pillars of digital India.

Source: Reprinted from https://it.assam.gov.in/portlet-level-1/digital-assam.

Digital Assam mission is mainly focusing on:

- Spread internet connectivity to the maximum reachable areas;
- Increasing the digital literacy rate in rural areas;
- Reaching government services and information to the citizens;
- Achieving better economy through digital platform;
- Bringing transparency and efficiency in governance through information technology.

Information and Communications Technology needs to be strengthened for the planning, implementation, and monitoring of various developmental schemes in the northeastern region. Major projects are also monitored through ICT-based technologies.

11.3.1 SCIENCE AND TECHNOLOGY INTERVENTION NORTHEASTERN REGION (STINER)

This innovative project was launched by the Ministry of Development of the Northeastern Region to provide technologies and knowledge to local communities for their skill development. This includes farmers, small-scale industries, and young entrepreneurs, among others, for social and economic development in the Northeastern Region. STINER has developed various important technologies that are useful for entrepreneurs, artisans, and farmers. These technologies include a handmade paper unit, paper plate and cup making machine, lamination machine, essential oil extraction plant, biogas plant, bio-fertilizer unit, plastic product designing facility, milk processing and packing plant, food testing facility, machining, and welding facility, among others.

11.4 SUGGESTIONS AND CONCLUSION

Northeastern India comprises eight states, and a significant portion of the region remains unexplored for business opportunities (Figure 11.2).

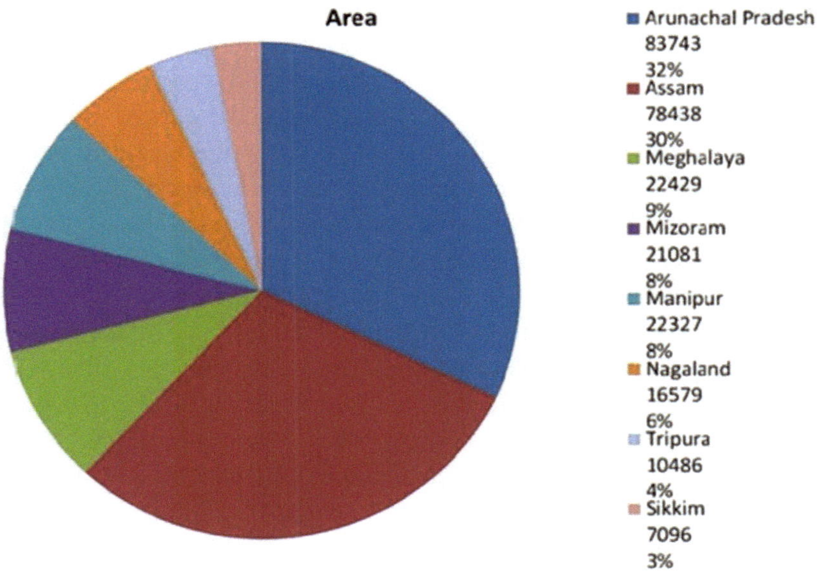

FIGURE 11.2 The area wise distributions of states.

The Northeast India is well known for its natural resources and ethnic communities with diverse cultures. This region has a significant tribal population with diverse customs, traditions, and languages. Despite having abundant natural resources, the region lags behind in various infrastructure developments due to its remote location and rugged terrain, making it difficult for investments. However, infrastructure development is gradually taking place. Among the northeastern states, Assam is the most developed state and is emerging as an economic and commercial hub in the region. Assam is also becoming a gateway to Southeast Asian countries, thus having the potential to play an important role in providing physical and digital infrastructure in the region. The literacy rate of the state is 73.18% as per the 2011 Census, slightly below the national average of 74%. The urban and rural literacy rates are 88.88% and 70.44% respectively. A significant number of young people, especially from towns, possess IT skills and are knowledgeable in the English language.

In terms of the infrastructure of Assam, the road network is the lifeline of the state. More state highways need to be developed to improve connectivity. It is important to upgrade rural roads to make them all-weather roads. Enhancing railway connectivity to remote locations is also essential. The establishment of more civil airports and air connectivity within the North-East region and other parts of India is necessary. The World War II airstrips in the North-East region can be utilized for developing new airports. Priority should be given to power sector reforms to ensure continuous energy supply. The development of telecommunication and robust wireless network connectivity is crucial in the present day.

The state of Assam is also gaining more importance in terms of infrastructure due to the special focus of the Government of India under the Look East Policy. Despite its remote geographical location, digital technology is also improving at a rapid pace, especially in urban areas. New technologies, supported by internet connectivity, are reaching the general population and creating new business opportunities for local communities, which will contribute to the socio-economic development of the region. The government should also consider incorporating local languages for implementing e-governance in remote areas. Many educational institutes in this region have also started to establish ICT-enabled labs and smart classrooms.

In the northeastern region of India, the major thrust area will be the development and improvement of civil infrastructure, transportation links, upgraded digital infrastructure, and the utilization of green energy such as solar and hydraulic energy. Additionally, equal importance will be given to creating an eco-friendly environment, managing e-waste, establishing IT parks, and supporting small-scale IT firms.

In the school curriculum structure, "ICT" should be included and online platforms could be created to make students digitally literate.

Promotional campaigns are essential for promoting 'Digital Northeast and Digital Assam.' The government should organize media campaigns, road shows, trade fairs, conclaves, and encourage local entrepreneurs to participate directly in promoting the region.

Digital payments by citizens, merchants, and traders must be encouraged. At the same time, the cyber security of these digital payment methods must be prioritized. Security and privacy of citizens' data should be mandated in all e-Governance programs.

A specific ICT-based training program may be conducted for students, women, physically challenged individuals, and economically disadvantaged people of the region. A centralized data-based help desk should be provided for citizens.

KEYWORDS

- **digital empowerment initiatives**
- **digital infrastructure**
- **ICT**
- **infrastructure development**
- **socio-economic impact**
- **sustainable growth**
- **tourism business**

REFERENCES

Annual Report, (2021–2022). Ministry of development of North Eastern region, Government of India.

CMIE, (2001). Infrastructure, Center for Monitoring Indian Economy, Mumbai. CMIE, Infrastructure, Various Issues, Center for Monitoring Indian Economy, Mumbai. Department of

Economics, Gauhati University and Indian Statistical Institute, Delhi Center (1987–88), The Structure of the Assam Economy (1982–83) and A Perspective Plan of Assam up to 2000 A.D.

Government of Assam, (1999). *Statistical Handbook Assam 1999.* Directorate of Economics and Statistics.

https://it.assam.gov.in (accessed on 11 January 2023).

https://mdoner.gov.in/about-ministry/annual-report (accessed on 11 January 2023).

https://pib.gov.in/Pressreleaseshare.aspx?PRID=1577192 (accessed on 11 January 2023).

IT Policy, (2017). *Information Technology and Electronic Policy Assam.*

Jain L.C. Committee (1990). *Report of the Committee on Clause Seven of Assam Accord.* Planning Commission, Government of India.

Sachdeva, G., (2000). *Economy of the North-East: Policy, Present Condition and Future Possibilities*, Konark Publishers, New Delhi.

Shukla S.P. Commission (1997). *Report of the High-Level Commission, Transforming the North-East-Tackling Backlogs in Basic Minimum Services, and Infrastructural Needs, Part I.* Planning Commission, Government of India.

Transport and Communication, (2022). Northeastern Council. Government of Assam.

CHAPTER 12

Firm Types and Design Processes: A Typological Analysis in the Indian Context

SANTANU ROY[1] and PARTHASARATHI BANERJEE[2]

[1]*CFAI Business School, The ICFAI University, Dehradun, India*

[2]*Indian Institute of Management, Raipur, Chhattisgarh, India*

ABSTRACT

Design plays a fundamental role in new product development and industrial innovation. Between the extremes of looking at an organization as a whole and the micro actors based processes, there remain internal structures and systems within the organization through which and by whose support the internal actions and processes shape up how the firm functions on ground. The present work reported is about these intermediate internal systemic structures as it looks at managerial processes pertaining to design competency in vogue in industrial firms located in the National Capital Region (NCR) of India and carries out a typological analysis of these firms based on these processes. We probe how the design processes are being practiced in a sample of 60 industrial firms of varying annual turnover and size in the automobiles, engineering, chemicals, pharmaceuticals, textiles, electrical and electronics sectors. We subsequently map out the relationships between: (i) firm type and design objectives; and (ii) firm type and sources of information. The results indicate that the majority of firms with a distinctive design department, demonstrate a better appreciation of what constitutes successful product innovation. Our results further

Sustainable Strategic Business Infrastructure Development and Contemporary Digital Practices in Industry 5.0. Indrajit Ghosal, Bikram Prasad, Nilanjan Ray, & Lalit Kumar Joshi (Eds.)

indicate the importance of design management and design processes that help boost innovation capabilities and capacities in industrial firms and the role of establishing networks in augmenting this initiative.

12.1 INTRODUCTION

The issue of organizational performance has remained the cornerstone of a large number of inquiries on business. Strategic management, in particular, has been investigating the performance of businesses and, in doing so, has often relied on theory or empirical-based predictions on firm-specific performances. However, this approach overlooks the emergence of strategy. To address this, the inquiry shifted towards the actions of micro-actors. Between the two extremes of organization as a whole and micro-actors-based processes, there are internal structures and systems within the organization that shape how the firm functions on the ground. This chapter focuses on these intermediate internal systemic structures and examines the managerial processes related to design competency in industrial firms located in the National Capital Region (NCR) of India. It also conducts a typological analysis of these firms based on these processes. Design plays a fundamental role in new product development and industrial innovation. We investigate different patterns of design engineering processes, including the formulation of design objectives and identification of information sources. We also examine whether designing agreements/arrangements with networked partners occur simultaneously with other parallel agreements, such as marketing, research and development (R&D), or service/training. We explore how the design processes are practiced in a sample of 60 industrial firms of varying annual turnover and size in the automobiles, engineering, chemicals, pharmaceuticals, textiles, electrical, and electronics sectors. Subsequently, we map out the relationships between: (i) firm type and design objectives; and (ii) firm type and sources of information. The results of the organizational processes-based typological analysis have significant implications for research and practice.

The significance of the conceptual construct of design continues to be recognized by various researchers (Armstrong, 1981; Freeman, 1992; Roy & Potter, 1993; Utterback, 1994; Peters, 1995; Fox, 1995; Rigelsford, 2000; Kidd & Thompson, 2000; Li & Boyle, 2009; Yin, Qin, & Holland, 2011; Maidin, Campbell, & Pei, 2012). Design may manifest itself in various forms or shapes and may relate to layouts of any work in progress or the final outcome of it, striving towards achieving certain well-defined objectives while considering the impediments that may arise (Wikipedia, 2022). Additionally, one can consider the component of styling.

The design process is closely linked to various elements of the innovation chain, including skills development, organizational development, vendor development, organizational strategy, and information management. It plays a critical role in establishing an organizational structure and its relationship with user activities. Busby & Williamson (2000) describe the results of a research program on measuring the performance of engineering design activities. Ellie (2014) discusses the advancements and innovations in virtual reality design. When it comes to complex engineering design concepts, such as aircraft or aircraft engine manufacturing, Murat, Marco, Mario, & James (2014) emphasize the importance of aligning customer needs with engineering characteristics. Gallagher (2014) highlights the significance of integrating design with engineering for enhanced effectiveness. Research and development (R&D) and quality control are encouraged in many economies (Biegelbauer & Borras, 2003; Borras, 2003; Malerba & Brusoni, 2007). It is time for design to receive similar recognition, particularly because it needs to be an integral part of innovation strategy and practice at both the firm and policy levels. Innovation-based approaches are crucial in such a scenario. However, there is limited acknowledgement in the existing literature, especially regarding developing economies. Macro-level indicators may indicate India's vast talent pool, but there are low indicators of innovation across all industries. How can managerial talent be better utilized to implement innovation? How can design-centered practices in engineering in Indian firms help them compete in global markets and meet growing demand at home? Our present research investigates firm-level design engineering practices in organizations located in NCR1, India. The motivation for this study is to critically understand the forms and modes of practice in design engineering in a developing economy like India, where innovation is crucial for the growth of technologically-driven firms. Managers in such firms need to enhance their capabilities for innovation to meet economic challenges in both the global and domestic marketplace.

The rest of the chapter is organized as follows: an overview of the literature, the research methodology adopted, presentation of study results, along with a discussion of key issues arising from our findings and their implications for research and practice.

12.2 AN OVERVIEW OF THE LITERATURE

12.2.1 DESIGN MANAGEMENT

The subject of design management is growing in importance in a dynamic business environment. Kalogerakis, Luthje, & Herstatt (2010) report a study that involved 18 German and Scandinavian design management organizations. There is a plethora of research that can be cited on the impact of effective design management on organizational performance and related aspects (Freeman, 1992; Roy & Potter, 1993; Utterback, 1994; Peters, 1995; Nystrom, 1985; Roy, 1994; Roy & Riedel, 1997; Feng & Kusiak, 2006; Sato, 2009; Design Council, 2007; Dekkers, Chang, & Kreutzfeldt, 2013; Hinckeldeyn, Dekkers, & Kreutzfeldt, 2015). Sato (2009) states, 'Design enables innovation, no matter what the business's strategy' (p. 44). The work carried out by Hong (2011) discusses how Apple creates well-designed products that allow it to gain an advantage across different sectors. Peter, Nicholas, and John (2014) investigate the fundamental information categories for engineering design. The issue of outsourcing the engineering design processes by a firm also assumes importance in this context (Zirpoli and Becker, 2011). Li and Qiu (2006) report an analysis of a large number of prior studies conducted with a view to integrating groups working across nations and regions, adopting rapidly evolving information technologies. In one of the earlier studies, it was observed that the long-standing neglect of design management had a profound negative impact on manufacturing firms based in the USA (Dixon and Duffey, 1990) in terms of their national and international standing.

12.2.2 INTEGRATING DESIGN AND THE DESIGN PROCESS

Studies have focused on sustainability and environmental issues concerning the design process (Boks & Stevels, 2007), benefits accrued from new product development and the success of new products (Cooper & Kleinschmidt, 1991), and how engineering firms in Thailand feature in this regard (Boonkanit & Kengpol, 2010). Gagnon, Leduc, & Savard (2012) refer to the concept of a sustainable design process. Kumar, Khan, & Gandhi (2011) present a literature review on design for maintainability. Regarding the appropriateness of systems adopted for manufacturing,

Caputo & Pelagagge (2008) assess how these are impacted by product-related features and design attributes.

In another study, considering the case of valves, Vinodh, Kamala, and Jayakrishna (2015) show how the adoption of a suitable design management approach improves important functional requirements such as manufacturability, adaptability, ease of operation, and strength and reliability. Similarly, there are other research studies that focus on integrating design and the design process (Luh, Ko, and Ma, 2011; Williams, Kannan, and Azarm, 2011; Keller, Eckert, and Clarkson, 2009).

12.2.3 UNDERSTANDING BUYERS, CUSTOMERS, AND NETWORKS

Understanding buyers, customers, and networks is critical in influencing and facilitating the transmission of know-how from national R&D laboratories to firms in need of this knowledge. It has been observed that technologies often perform well in bench-scale or laboratory-scale settings, but may experience lower performance levels in commercial-scale operations, despite efforts to find innovative solutions to these challenges (Roy & Mohapatra, 2002; Roy, 2004, 2006, 2009; Roy & Banerjee, 2007; Von, 1986).

A firm that has achieved a thorough appreciation of what buyers and customers are looking for and has adopted a workable network approach to link the channel partners can engage in an extension effort, seeking ways to mobilize seamless sharing of information across channel partners to bridge the gap between customer expectations and the ways to achieve that end by designing matching products and processes (McCardle, 1985; Brown & Utterback, 1985; Spitas, 2011).

In the Indian scenario, the industrial and trade protection regimes of the past have hindered the recognition and utilization of design in engineering and innovation for a long time. According to Frankena (1974), the designs of many engineering goods produced in India during the 1960s were not efficient; in fact, they were technically obsolete or economically inefficient. However, the process of economic liberalization initiated in 1991 and the subsequent opening up of the Indian economy, relaxation of ownership controls, higher levels of foreign direct investment (FDI), success with information technologies, and exposure to greater competition have transformed the industrial landscape. The concept of design has gained prominence with some notable success stories (Toloken, 2010; Ganguli, 2010), the enhancement of existing facilities (Yeates, 2002), and its role as an enabler of agile manufacturing (Vinodh & Kuttalingam, 2011). According to Tippu (2008), the success of engineering design in India is based on three fundamental factors: a talented pool of individuals, industry-academic cooperation, and government support.

12.3 METHODS AND DATA

For the present study, a convenience sample of 60 firms located in NCR, India, operating in the domains of engineering, chemicals, automobiles, and electronics, and having different turnover, were selected. Based on theoretical concepts of the design and innovation processes, as well as field investigations carried out, a questionnaire was developed and used in the study. The responses were thoroughly scrutinized to investigate aspects of the various design processes adopted by the firms and the design-related outcomes for innovation.

The firms that participated in the survey belonged to two sets – Set 1 (firms with their own Design Cells) and Set 2 (firms without their own Design Cells). Based on our literature review, we extract the value of a strategic approach to design management for innovation, focusing on the significance of four essential dimensions of design-led strategy in engineering firms:

- The acquisition of assets (machinery) in firms;
- The maintenance of information in databases;
- Ranked design objectives; and
- Network management.

The results of our findings, shown in the following sections, are organized to align with the four dimensions of design-based strategy.

12.4 RESULTS

12.4.1 PHYSICAL ASSETS

The aspects of the physical assets possessed by the firms in terms of asset value of firm machinery as a percentage of firm gross turnover are presented in Table 12.1. It may be noted from Table 12.1 that the Asset Values of Firm Machinery as a percentage of Firm Gross Turnover for firms possessing a Design Cell of their own (Set 1) are substantially lower than the figures for firms that do not possess any Design Cell of their own (Set 2). Further scrutiny reveals that for firms having a turnover of less than INR 60 million, large firms (a turnover of INR 300–600 million) and firms having a turnover figure of more than INR 600 million, this difference is more noticeable, 81 INR=1 USD.

TABLE 12.1 Asset Value of Firm Machinery as a Percentage of Firm Gross Turnover

Firm Gross Turnover (in Million Indian Rupees, INR)*2	Set 1	Set 2
Less than or equal to 60	14.78	34.76
60–299	19.01	22.54
300–600	16.54	38.11
600 or higher	13.47	40.03
Mean	15.95	33.86

*81 INR = 1 USD.

Aspects of Sets 1 and 2 firms are subject to further scrutiny as detailed in the succeeding subsections.

12.4.2 KNOWLEDGE DATABASES

The extent of knowledge databases maintained by the different sets of firms is presented in Table 12.2. It should be noted that all types of firms maintain knowledge databases of some value, except for firms in the chemicals domain. Unsurprisingly, firms in the electronics sector domain do not maintain formulation databases, but they do maintain the highest level of marketing databases. Firms in the automobiles domain maintain the highest level of manufacturing and inventory databases. Therefore, the maintenance of inventory databases is common across industries within the engineering domain, but not in the chemicals and electronics domains. Importantly, firms operating in the automobiles and engineering domains maintain marketing databases at a lower level compared to firms in other sectors.

TABLE 12.2 Database Maintenance

Firm Type	Percentage of Firms (Engineering Databases)	Percentage of Firms (Formulation Databases)	Percentage of Firms (Manufacturing Databases)	Percentage of Firms (Inventory Databases)	Percentage of Firms (Marketing Databases)
Automobiles	68.54	23.12	76.88	78.02	34.23
Engineering	55.64	28.06	53.98	46.07	28.02
Chemicals	29.11	16.02	43.05	29.04	43.18
Electronics	67.28	0.0	56.12	23.11	45.98

12.4.3 FIRST CHOICE FIRM DESIGN OBJECTIVES

Table 12.3 presents the design objectives for the first-choice firm. For firms that have their own Design Cells (Set 1), customer satisfaction is of utmost importance, with their needs and demands taking precedence. Interestingly, while Reliability is still important, it is ranked lower in terms of objectives. It is worth noting that even for firms without a separate Design Cell (Set 2), customer needs and choices remain the top priority, followed by other secondary objectives that are maintained at the same level across different industrial domains. The factors of Optimality and Reliability also hold some significance.

TABLE 12.3 First Choice Firm Design Objectives

Firm Set	Meeting Unique Customer Demands	Minimum Consumption of Materials/ Resources	Surpass the Features of Competitors' Products	Ease of Manu-facture	Ergonomics and Ease of Operation	Opti-mality	Relia-bility
Set 1	79.8	0	0	0	7.3	0	14.6
Set 2	47.1	13.2	13.2	0	13.2	7.4	7.4

12.4.4 SECOND CHOICE FIRM DESIGN OBJECTIVES

The design objectives of the second-choice firm are detailed in Table 12.4. Among Set 1 firms, the dominant objective is to minimize material and resource consumption. For Set 2 firms, the top priority is reliability. Other significant factors include ease of manufacturing and surpassing the features of competitors' products. Another noteworthy objective is the minimum consumption of materials and resources.

TABLE 12.4 Second Choice Firm Design Objectives

Firm Set	Meeting Unique Customer Demands	Minimum Consumption of Materials/ Resources	Surpass the Features of Competitors' Products	Ease of Manu-facture	Ergonomics and Ease of Operation	Opti-mality	Relia-bility
Set 1	7.8	29.4	14.9	7.8	20.9	14.4	7.6
Set 2	0	10.8	21.9	21.9	0	0	44.8

12.4.5 DESIGN AND INDUSTRIAL INNOVATION: SIGNIFICANCE OF NETWORKS

Tables 12.5–12.8 focus on aspects of design and innovation and the significant roles networks play in this scenario. Table 12.5 provides details on the first-choice sources of information, while Table 12.6 provides details on the second-choice sources of information.

The Marketing Team, followed by Dealers, are the top choices as sources of information for Set 1 firms. For Set 2 firms, the distribution is more widespread, with the Marketing Team still dominating ranking way above vendors/suppliers, and the common position held by Dealers, Services Team, and Joint Venture/Network Partner, all figuring at the same level. Focusing on the integration of marketing and engineering, Luo (2011) writes, "When designing a line of consumer durable products, firms need to account for not only the interrelationships between consumer preferences and engineering feasibility/restrictions in the design of each product but also the revenue and cost interactions across the products in the product line. Furthermore, to forecast the revenue from a product line accurately, it is critical for firms to account for the strategic reactions from competing manufacturers and the retailer when the new product line enters the market" (p. 129).

TABLE 12.5 First Choice Sources of Information

Firm Set	Marketing Team	Dealers	Market Survey	Vendors	Service Team	Joint Venture/ Network Partner
Set 1	61.3	32.8	7.4	0	0	0
Set 2	38.2	13.7	4.1	19.2	13.7	13.7

TABLE 12.6 Second Choice Sources of Information

Firm Set	Marketing Team	Dealers	Market Survey	Vendors	Service Team	Joint Venture/ Network Partner
Set 1	31.1	24.2	14.7	9.9	14.6	7.9
Set 2	13.9	27.3	13.9	9.1	22.7	13.5

The details of parallel agreements that complement a design agreement with a networked partner are displayed in Table 12.7, focusing on the significance of developing networks to augment the design processes. For firms possessing their own Design Cells (Set 1), the design agreements co-evolve with Marketing, Servicing/Training, R&D, and Design Engineering arrangements. For Set 2 firms, all these arrangements (except the one for Marketing) have some bearing on their work, but the arrangements tend to be made much less frequently than those established for Set 1 firms.

TABLE 12.7 Parallel Agreements Along with Establishing a Design Agreement with a Networked Partner

Firm Set	Marketing Agreement	Service/Training Agreement	R&D Agreement	Design Engineering Agreement
Set 1	36.9	36.9	42.6	42.6
Set 2	5.4	15.4	15.4	15.4

The details of who generally provide the design specifications and engineer the prototypes for the two categories of firms are presented in Table 12.8. A perusal of Table 12.8 reveals that, excluding joint venture partners and technical collaborators, all the other network partners of the firm contribute to the design process and engineering of prototypes for Set 1 firms. However, these firms primarily rely on their in-house design departments to carry out such tasks most of the time, although the other group collaborators have a fair role to play in this task. The percentages are much lower for Set 2 firms.

12.5 DISCUSSION, IMPLICATIONS, AND CONCLUDING OBSERVATIONS

The study focuses primarily on the indicators of industrial innovation. We observe how the baseline operations of firms with and without separate design departments impact their approach to innovation. For instance, Category 1 firms, which have separate design departments and are considered design innovators in our sample, rely less on acquired assets compared to their Category 2 counterparts. Design Innovators utilize internal design teams to provide design specifications and engineers to create prototypes. Meeting unique customer demands remains their primary design objective, and these firms depend on their own marketing teams and dealer firms as their primary source of information. The results indicate that the majority of firms with a distinct design department demonstrate a better understanding of what constitutes successful product innovation. This is achieved through forging partnerships and developing arrangements for marketing, service, or R&D tasks.

TABLE 12.8 Provider of Design Specifications and Engineering Prototypes

Firm Set	Joint Venture Partner/ Technical Collaborators	Own Design Team/Other In-House Team	Business Associate/ Partner	Engineer-ing Consul-tant	R&D Consul-tant	Software Consul-tant	Out-sider
Set 1	8.8	75.4	25.2	25.2	16.9	16.9	32.8
Set 2	19.3	38.6	0.2	5.1	5.1	0.2	19.4

The results have resonance in a few other studies carried out and reflected in the literature. The study carried out by Luh, Ko, & Ma (2011) in a medium-sized aerospace company points out the importance of documentation and problem-solving. Our findings also reflect what was noted in earlier work carried out by other researchers (Williams, Kannan, & Azarm, 2011; Keller, Eckert, & Clarkson, 2009).

Empirical studies (Phelps, 2010) have also addressed issues such as innovation, resources, and capabilities (Arikan & McGahan, 2010), focusing on the organization as a whole. This aligns with the research conducted by Salter & Gann (2003), which explores the sources of ideas for innovation in engineering design. It demonstrates that engineering designers involved in complex, non-routine design processes heavily rely on face-to-face conversations with other designers to solve problems and develop new innovative ideas. Our study results are consistent with the aforementioned findings. Our results further indicate the importance of design management and design processes that help boost innovation capabilities and capacities in industrial firms, as well as the role of establishing networks in augmenting this initiative. Piirainen, Kolfschoten, and Lukosch (2012) focus on an improved understanding of collaborative design, which is also the focus of attention of Crabtree, Fox, and Baid (1997), and the specific management problems associated with it. Gendron, Pourroy, Carron, and Marty (2012) emphasize the need for creating collaborative platforms for sharing knowledge in engineering design and probe the issue of defining adequate indicators for such activities.

12.6 LIMITATIONS

Our cross-sectional study provides a snapshot, which can make generalization difficult. This issue is compounded by the problem of data availability, particularly reliable and consistently maintained data. This is understandable in the context of development, where new approaches to innovation management are beginning to take root. However, our findings do provide adequate insights into the design engineering practices of firms in such a context, as well as in terms of the closeness of the findings to other empirical work in different environments, enabling robust comparisons to be drawn. It is important to note that our sample size is limited, given the exploratory nature of the study.

KEYWORDS

- **data availability**
- **design management**
- **design processes**
- **firm types**
- **Indian context**
- **NCR**
- **typological analysis**

REFERENCES

Arikan, A. M., & McGahan, A. M., (2010). The development of capabilities in new firms. *Strategic Management Journal*, *31*(1), 1–18.

Armstrong, I., (1981). Design for production. *Industrial Management & Data Systems, 81* (9, 10), 2–4.
Biegelbauer, P., & Borras, S., (2003). *Innovation Policies of Europe and US: The New Agenda*. Ashgata Publishing, England.
Boks, C., & Stevels, A., (2007). Essential perspectives for design for environment: Experiences from the electronics industry. *International Journal of Production Research, 45*(18, 19), 4021–4039.
Boonkanit, P., & Kengpol, A., (2010). The development and application of a decision support methodology for product eco-design: A study of engineering firms in Thailand. *International Journal of Management, 27*(1), 185–199.
Borras, S., (2003). *Innovation Policy of the EU: From Government to Governance*. Edward Elgar, UK.
Brown, J. W., & Utterback, J. M., (1985). Uncertainty and technical communication patterns. *Management Science, 31*(3), 301–311.
Busby, J. S., & Williamson, A., (2000). The appropriate use of performance measurement in non-production activity: The case of engineering design. *International Journal of Operations and Production Management,20*(3), 336–358.
Caputo, A. C., & Pelagagge, P. M., (2008). Effect of product design on assembly lines performances: A concurrent engineering approach. *Industrial Management & Data Systems, 108*(6), 726–749.
Cooper, R. G., & Kleinschmidt, E. J., (1991). New products: What separates winners from losers? In: Henry, J., & Walker, D., (eds.), *Managing Innovation*. Sage, London.
Crabtree, R. A., Fox, M. S., & Baid, N. K., (1997). Case studies of coordination activities and problems in collaborative design. *Research in Engineering Design, 9*, 70–84.
Dekkers, R., Chang, C., & Kreutzfeldt, J., (2013). The interface between product design and engineering and manufacturing: A review of the literature and empirical evidence. *International Journal of Production Economics, 144*(1), 316–333.
Design Council, (2007). Designing demand: The importance of design. *Strategic Direction, 23*(7), 30–32.
Dixon, J. R., & Duffey, M. R., (1990). The neglect of engineering design. *California Management Review, 32*(2), 9–23.
Ellie, Z., (2014). Designing a new reality. *Engineer, 296*(7850), 37, 38.
Feng, J. C. X., & Kusiak, A., (2006). Data mining applications in engineering design, manufacturing, and logistics: Guest editorial. *International Journal of Production Research, 44*(14, 15), 2689–2694.
Fox, J., (1995). Managing design for a quality product. *Assembly Automation, 15*(1), 12–14.
Frankena, M., (1974). The industrial and trade control regime and product designs in India. *Economic Development & Cultural Change, 22*(2), 249–264.
Freeman, C., (1992). *The Economics of Hope: Essays on Technical Change, Economic Growth and the Environment*. Frances Pinter, London.
Gagnon, B., Leduc, R., & Savard, L., (2012). From a conventional to a sustainable engineering design process: Different shades of sustainability. *Journal of Engineering Design, 23*(1), 49–74.
Gallagher, M., (2014). Integrating design with engineering. *Research Technology Management, 57*(3), 64.

Ganguli, R., (2010). Rotorcraft research in India: Recent developments. *Aircraft Engineering and Aerospace Technology, 82*(5), 305–319.
Gendron, E., Pourroy, F., Carron, T., & Marty, J., (2012). Towards a structured approach to the definition of indicators for collaborative activities in engineering design. *Journal of Engineering Design,23*(3), 195–216.
Hinckeldeyn, J., Dekkers, R., & Kreutzfeldt, J., (2015). Productivity of product design and engineering processes: Unexplored territory for production management techniques? *International Journal of Operations and Production Management, 34*(4), 458–486.
Hong, J., (2011). Matters of design: Part II. *Communications of the ACM, 54*(4), 10, 11.
Kalogerakis, K., Luthje, C., &Herstatt, C., (2010). Developing innovations based on analogies: Experience from design and engineering consultants. *Journal of Product Innovation Management, 27*, 418–436.
Keller, R., Eckert, C. M., & Clarkson, P. J., (2009). Using an engineering change methodology to support conceptual design. *Journal of Engineering Design, 20*(6), 571–587.
Kidd, M. W., & Thompson, G., (2000). Engineering design change management. *Integrated Manufacturing Systems, 11*(1), 74–77.
Kumar, S., Khan, I. A., & Gandhi, O. P., (2011). A kaleidoscopic view of psychology in design for maintainability in mechanical systems: A review. *Journal of Engineering, Design and Technology, 9*(3), 347–370.
Li, S., & Boyle, I. M., (2009). Engineering design: Perspectives, challenges, and recent advances. *Journal of Engineering Design, 20*(1), 7–19.
Li, W. D., & Qiu, Z. M., (2006). State-of-art technologies and methodologies for collaborative product development systems. *International Journal of Production Research, 44*(13), 2525–2559.
Luh, D., Ko, Y., & Ma, C., (2011). A structural matrix-based modelling for designing product variety. *Journal of Engineering Design, 22*(1), 1–29.
Luo, L., (2011). Product line design for consumer durables: An integrated marketing and engineering approach. *Journal of Marketing Research, 48*(1), 128–139.
Maidin, S. B., Campbell, I., & Pei, E., (2012). Development of a design feature database to support design for additive manufacturing. *Assembly Automation, 32*(3), 235–244.
Malerba, F., & Brusoni, S., (2007). *Perspective on Innovation*. Cambridge University Press, London.
McCardle, K. F., (1985). Information acquisition and the adoption of new technology. *Management Science, 31*(11), 1372–1389.
Murat, H. E., Marco, B., Mario, K., & James, C., (2014). Mapping customer needs to engineering characteristics: An aerospace perspective for conceptual design. *Journal of Engineering Design, 25*(1–3), 64–87.
Nystrom, H., (1985). Company strategies for designing and marketing new products in the electrotechnical industry. In: Longdon, R., & Rothwell, R., (eds.), *Design and Innovation: Policy and Management* (pp. 18–26). Frances Pinter, Frances Pinter, London.
Peter, H., Nicholas, H. M. C., & John, C. P., (2014). Core information categories for engineering design: Contrasting empirical studies with a review of integrated models. *Journal of Engineering Design, 25*(1–3), 88–124.
Peters, T., (1995). Pundit of passion. *Design* (pp. 18, 19). Summer.

Phelps, C. C., (2010). A longitudinal study of the influence of alliance network structure and composition on firm exploratory innovation. *Academy of Management Journal, 53*(4), 890–913.

Piirainen, K. A., Kolfschoten, G. L., & Lukosch, S., (2012). The joint struggle of complex engineering: A study of the challenges of collaborative design. *International Journal of Information Technology and Decision Making, 11*(6), 1087–1125.

Rigelsford, J., (2000). Design methods in engineering and product design. *Assembly Automation, 20*(1), 86, 87.

Roy, R., & Potter, S., (1993). The commercial impacts of investments in design. *Design Studies, 14*(2), 171–193.

Roy, R., & Riedel, J. C., (1997). The role of design and innovation in competitive product development. *Second European Academy of Design Conference: Contextual Design – Design in Contexts*. Stockholm, Sweden.

Roy, R., (1994). Can the benefits of good design be quantifies? *Design Management Journal, 5*(2), 9–17.

Roy, S., & Banerjee, P., (2007). Developing regional clusters in India: The role of national laboratories. *International Journal of Technology Management and Sustainable Development, 6*(3), 193–210.

Roy, S., & Mohapatra, P. K. J., (2002). Regional specialization for technological innovation in R&D laboratories: A strategic perspective. *Artificial Intelligence and Society, 16*, 100–111.

Roy, S., (2004). Technology transfer from national R&D laboratories and the development of regional industrial clusters in India. *Industry and Higher Education, 18*(1), 39–45.

Roy, S., (2006). Networking as a strategy for technology transfer and commercialization from R&D laboratories. *Industry and Higher Education, 20*(2), 123–133.

Roy, S., (2009). Technology transfer from R&D centers in India and development of industrial clusters. In: Manimala, M. J., Mitra, J., & Singh, V., (eds.), *Enterprise Support Systems: An International Perspective* (pp. 217–227). ISBN: 978-81-7829-927-3, Sage, New Delhi.

Salter, A., & Gann, D., (2003). Sources of ideas for innovation in engineering design. *Research Policy, 32*(8), 1309–1324.

Sato, S., (2009). Beyond good: Great innovations through design. *Journal of Business Strategy, 30*(2, 3), 40–49.

Spitas, C., (2011). Analysis of systematic engineering design paradigms in industrial practice: A survey. *Journal of Engineering Design, 22*(6), 427–445.

Tippu, S., (2008). *COO on India Design: Boom* (Vol. 1515, p. 22). Electronic Engineering Times.

Toloken, S., (2010). *Indian Firm Designs WPC Toilet* (Vol. 21, No. 41, p. 12). Plastics News.

Utterback, J. M., (1994). *Mastering the Dynamics of Innovation*. Harvard Business School Press, Boston, MA.

Vinodh, S., & Kuttalingam, D., (2011). Computer aided design and engineering as enablers of agile manufacturing: A case study in an Indian manufacturing organization. *Journal of Manufacturing Technology Management, 22*(3), 405–418.

Vinodh, S., Kamala, V., & Jayakrishna, K., (2015). Application of fuzzy axiomatic design methodology for selection of design alternatives. *Journal of Engineering, Design and Technology, 13*(1), 2–22.

Von, H. E., (1986). Lead users: A source of novel product concepts. *Management Science, 32*(7), 791–805.

Wikipedia, (2022). *Design.* https://en.wikipedia.org/wiki/Design (accessed on 11 January 2023).

Williams, N., Kannan, P. K., & Azarm, S., (2011). Retail channel structure impact on strategic engineering product design. *Management Science, 57*(5), 897–914.

Yeates, H., (2002). Test firm to double staff at Indian design facility. *Electronics Weekly, 2055*, 10.

Yin, Y., Qin, S., & Holland, R., (2011). Development of a design performance measurement matrix for improving collaborative design during a design process. *International Journal of Productivity and Performance Management, 60*(2), 152–184.

Zirpoli, F., & Becker, M. C., (2011). The limits of design and engineering outsourcing: Performance integration and unfulfilled promises of modularity. *R&D Management, 41*(1), 21–43.

CHAPTER 13

Wearable Technology: Developments, Business Models, and Current Challenges

PRIYANKA SINGH and HARIT KUMAR

Pranveer Singh Institute of Technology, Kanpur-Agra-Delhi National Highway-2, Bhauti, Kanpur, Uttar Pradesh, India

ABSTRACT

Technology is undergoing a fundamental transformation as a result of the emergence of countless interconnected "things" and their integration into human daily life. One of the key technological solutions supporting this integration is wearable devices, which can gather the vast contextual information generated by these technologies to provide a truly personalized experience. The main objective of this research is to provide a historical context for wearable technology and an up-to-date analysis of the current wearable market. Additionally, the study covers various other topics related to wearable technology, including a comprehensive and multidimensional categorization of wearable devices based on different variables, discussions on wireless communication technologies, designs, information management aspects, business positioning, and more. The study concludes by highlighting major challenges and presenting both current and potential solutions.

13.1 INTRODUCTION

Wearable technology devices are accessories with built-in digital technology, such as wearable devices that attach directly to the skin

Sustainable Strategic Business Infrastructure Development and Contemporary Digital Practices in Industry 5.0. Indrajit Ghosal, Bikram Prasad, Nilanjan Ray, & Lalit Kumar Joshi (Eds.)

(Godfrey et al., 2018). Microchips, sensor systems, portable and wireless gadgets connected to servers collect information about human habits in wearable devices, which is part of the Internet of Things (IoT) (Chawla, 2020). Technological interventions, such as devices that integrate customers and service providers like insurers and programmers, are expected to change the competitive landscape in the respective industry (Porter & Heppelmann, 2014). With real-time data about a person's lifestyle provided by wearable technology, the insurance industry can offer personalized solutions and products to its consumers (Porter & Heppelmann, 2014). Fitness and wellness devices are the two main categories into which wearable technology can be divided (Godfrey et al., 2018). The first category of devices focuses on providing individuals with real-time updates on their exercise routines through streamlined target mapping frameworks (Adams, 2018). The second category enables the provision of specific intervention programs for patients with conditions that require ongoing monitoring (Rastogi et al., 2008). Descriptions of wearable technology include digital watches, cameras, armbands, earpieces, sensors, activity trackers, skin patches, walkie-talkies, cell phones, clothing, eyewear, and monitoring devices (Godfrey et al., 2018). Nicotine patches and calorie counters help people quit smoking and improve their diet, respectively. Some of the most well-known brands of wearable technology are Smartwatch, Fitbit, Google Fit, Samsung Gear, and Jawbone (Gao et al., 2015).

13.2 BACKGROUND

Wearable technologies, such as smart watches, fitness trackers, wellness gadgets, and fitness monitors, have become increasingly popular among young customers in recent years. Consumers are becoming more aware of wearable devices and software that can make daily life easier. The growing demand for wearable technology in India is driven by increased disposable income and consumer awareness.

Wearable device usage is accelerating across the country as 4G/5G and Wi-Fi networks become more widely available. The fastest-growing economies in the world are adopting 4G/5G networks and are likely to boost the market for wearable technology and devices.

Some of the key participants in the wearable device market include Apple India Pvt. Ltd., Samsung India Electronics Pvt. Ltd., Microsoft

Corporation (India) Pvt. Ltd., Nike India Pvt. Ltd., Intel Technology India Pvt. Ltd., and Google India Pvt. Ltd.

Global companies such as Samsung, Google, Apple, and Intel are investing billions in research and development for wearable innovations due to their numerous potential uses. Forecasts indicate that the wearable technology market will grow from its current $20 billion in 2015 to reach $70 billion by 2025, thanks to the aforementioned investment (Raj & Ha-Brookshire, 2016).

13.2.1 WHAT ARE WEARABLE TECHNOLOGIES?

Wearable technologies lack a consistent definition in the academic literature. However, there are a few terms that are essentially synonymous with wearable technologies and have closely related meanings. Expressions like "wearable gadgets," "wearable computers," and "wearable electronics" all refer to the same thing. Except for "wearable computer," all of these phrases have similar meanings and can be used interchangeably, as indicated by their interpretations. According to Dunne (2004), there is a broader category of wearable technology that includes wearable computer systems, but the distinction between wearable technology and wearable computer is unclear at best (Profita, 2011). A wearable technology is defined as a technology-enabled computing machine that takes and processes data. It is commonly used as a form of body modification or adornment. This gadget can operate independently or connect to a user's smartphone to facilitate more extensive communication. A wearable product can be something worn on the body (such as an intelligent bandage), around the body (like a watch or a headband), or even inside the body (such as a sub-dermal recognition detector or a detector linked to tracing cardiac abnormalities through heartbeats) (Kurwa et al., 2008).

In the path to unveil the concept and determinants of wearable technology, many domains have been explored, demonstrating a variety of instruments that provide an overview of the distant wearable technology model and techniques originating in organizations. However, in recent years, there has been a severe lack of research in wearable technology studies determining the model, characteristics, and usability approaches to inform academics and professionals. As a result, this current writing addresses the current state of wearable technology research and the most often applied wearable technology tools and approaches (e.g., physiological and neurophysiologic) in the business sector over the previous two decades. The following is a list of research investigations that our literature review sought to address:

- **RQ1:** Understanding the key business domain and usability/types of wearable technology.
- **RQ2:** Which PRISMA framework selected noteworthy contributions of the most cited and productive articles, journals, countries, scholars, and keywords in wearable technologies?
- **RQ3:** What is the key component for identification on the wearable technology modeling canvas?

Answering these inquiries will provide future generations with a solid foundation in the methods and literature currently employed in descriptive WT research. The goal of this writing is to draw the most comprehensive and effective conclusion possible. Additionally, study topics are often examined in depth according to sub-domains.

13.3 PLAN OF ACTION

By conducting an in-depth literature review to assess the level of study in various publications and identify different tools and techniques of WM, this study provides an answer to RQ1.

The current study utilized bibliometric analysis to address RQ2 regarding the latest advancements in the field of wearable technology. By analyzing publications spanning two decades, the bibliographic/bibliometric analysis revealed global trends, including the most productive journals, contributors, countries, and institutions, as well as the most cited articles in relevant topic areas. The research objectives were achieved through a comprehensive search of the Scopus database for articles related to the wearable technologies initiative. The investigation covered the period from 2000 to 2020. Scopus, with 24,018 indexed titles as of November 22, is a highly recommended bibliographic database that encompasses a wide range of high-quality scholarly publications. This study contributes to the understanding of the qualitative aspects of the collected data, as well as the conclusions, limitations, and recommendations.

This approach will enable the researcher to better comprehend the progress of the WT field. The finding is supported by searching for articles about "Wearable Technology" in the electronic database.

Using citation analysis, the significant advantages were laid out in considerable detail (as the leading contributor in this field). The most popular and influential countries, researchers, journals, keywords, and their ranking with effectiveness were also examined. Bibliometric analysis was performed on the data with the help of VOSviewer Software.

In response to the third research question, we applied the "Preferred Reporting Items for Systematic Reviews and Meta-Analyses (PRISMA)" methodology developed by Moher et al. (2015). This methodology is widely used as a reporting guideline for systematic reviews. A total of 246 articles were included for content analysis in this chapter, following various inclusion and exclusion criteria. These studies covered the fields of business, psychology, arts, humanities, and social sciences, totaling 1,644 articles. The number was further narrowed down to 775 articles at the publication stage. Subsequently, the language was restricted to English only, resulting in 742 articles, all of which were open access 246.

Inclusion and exclusion criteria:

> *"=TITLE-ABS-KEY-AUTH ("wearable technology") AND (LIMIT-TO (SUBJAREA, "SOCI") OR LIMIT-TO (SUBJAREA, "BUSI") OR LIMIT-TO (SUBJAREA, "PSYC") OR LIMIT-TO (SUBJAREA, "ARTS")) AND (LIMIT-TO (DOCTYPE, "ar")) AND (LIMIT-TO (LANGUAGE, "English")) AND (LIMIT-TO (OA, "all")"*

13.3.1 RQ1: KEY BUSINESS DOMAIN AND USABILITY/TYPES OF WEARABLE TECHNOLOGY

Literature reviews reveal that the zonal of wearable technologies has as widespread and scope ranging from communication, textile, sports, medical, fashion accessories as so on, and body placement from head to toe. Thus, chapter represent details domain analysis on used and applicability listed as (Figure 13.1):

- The ability to communicate (C) opens the door to sharing data with neighboring nodes and/or a cloud service located elsewhere.
- Capability for input and control (CI) is a broad category that encompasses a range of devices, from smart buttons to advanced motion capture detection hardware. The main objective of this team is to enhance the conventional human-computer interface by focusing on device usability, while also ensuring a compact form factor.

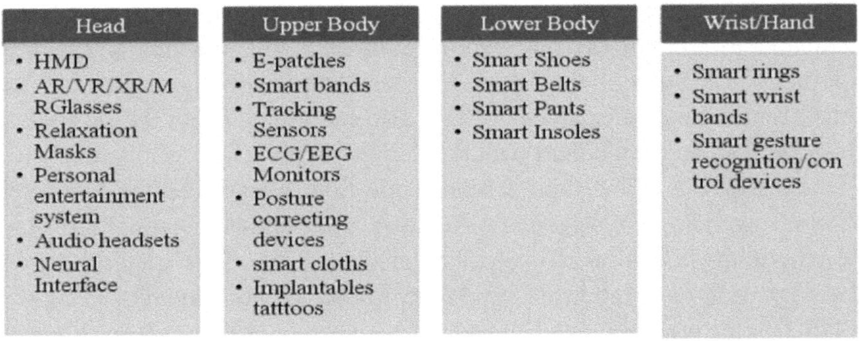

FIGURE 13.1 Illustrates about the various body fragmentation of wearable devices.

- The goal of the Learning and Professional Sports Domain (ESD) is to enhance teaching and coaching through the use of observers.
- Capabilities for amusement, gaming, and relaxation, such as audio systems and personal entertainment displays, are all examples of technologies that enhance the perceptual experience. Technologies that improve the perceptual experience include features for entertainment, gaming, and leisure, like audio systems and personal entertainment displays. Moreover, wearable cameras, IoT devices, AR/VR, and teleconsultation can improve the manner in which users are typically provided with information.
- Medical/healthcare features (HM). Similar devices like electrocardiogram (ECG) and electroencephalogram (EEG) monitors, relaxation stress relieving aids, neural interactions, prosthetic limbs, and so on fall under the umbrella of "medical devices," but must undergo extensive research and development (R&D) to earn this status.
- To perform geo-location mapping (GM), a global navigation satellite system (GNSS) or another form of on-board wireless communication device is required. The concept here is similar to both the node's awareness of its physical location and, if needed, the device's remote localization.
- The functionality for sending alerts (N) can range from something as basic as a vibration to something as advanced as an augmented reality add-on. Similarly, the sensing functionality can be supported by practically any personal device connected to the cloud, either directly or through the gateway.

- Functionality (O) is to send information from the user's personal ecosystem to the user and/or other individuals in the immediate area, using a variety of visual, auditory, and/or haptically enabled devices.
- Functions related to safety and security (S) include things like personal safety gadgets and emergency helpers.
- Keeping an eye on performance: This capability can be implemented with minimal effort and cost. Any gadget with an inertial sensor built in can perform basic sensing functions, such as heart rate monitors, blood pressure displays, and brainwave recorders, for medical gadgets used in health and fitness.
- Intelligent collars/neckbands, bark collars, high-tech clothing, and other animal wearable devices (PF) are all included in this category.

Classification and categories of wearable technology application are given in Figure 13.2.

13.3.2 THE PRISMA FRAMEWORK HAS SELECTED NOTEWORTHY CONTRIBUTIONS FROM THE MOST CITED AND PRODUCTIVE ARTICLES, JOURNALS, COUNTRIES, SCHOLARS, AND KEYWORDS IN THE FIELD OF WEARABLE TECHNOLOGIES

The key objective of this section of the chapter is to delve deep into the academic landscape of wearable technology. The outcomes of academic research in this field greatly contribute to the advancement of wearable technology. In terms of citations, the top contributors in this domain are Chuah, Krey, Lade, Nguyen, Ramayah, and Rauschnabel with 276 citations, followed by Tom Dieck M.C. with 222 citations, and Moore with 220 citations, as illustrated in Table 13.1.

The flag bearer of the most cited country in the wearable technology field is represented in the United States with 1,814 citations and the United Kingdom with 1,308 contributions. Subsequently, the ranking belongs to the Netherlands, Italy, Spain, and Germany with 393, 277, 273, and 159 citations, respectively, as mentioned in Table 13.2. Figure 13.1 represents the visualized linkage among the nations' research amalgamation.

Figure 13.3 represents a comprehensive analysis of keywords related to wearable technologies and the outcomes derived from academic research in this field. The analysis covers various aspects such as wearable physical activity, machine learning, wearable devices, self-tracking, IoT, technology, quantified self, health, big data, augmented reality, wearable sensors, social media, privacy, and personal informatics. Table 13.3 provides further details by indicating the occurrence and total link strength of these keywords.

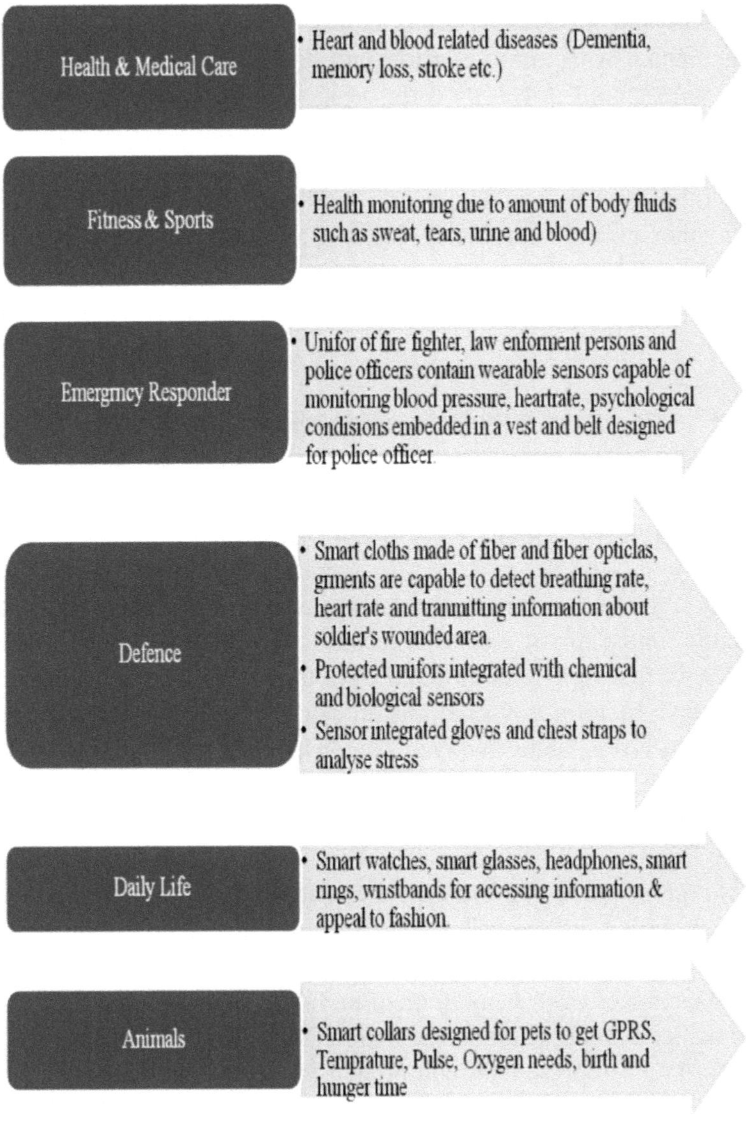

FIGURE 13.2 Business classification of wearable technology and area coverage.

TABLE 13.1 Most Cited Author's in Field of Wearable Technology

SL. No.	Author	Documents	Citations	Total Link Strength
1.	Chuah, S. H.-W.	1	276	5
2.	Krey, N.	1	276	5
3.	Lade, S.	1	276	5
4.	Nguyen, B.	1	276	5
5.	Ramayah, T.	1	276	5
6.	Rauschnabel, P. A.	1	276	5
7.	Tom Dieck, M. C.	2	222	0
8.	Moore, P.	3	220	1
9.	Robinson, A.	1	162	1
10.	Bower, M.	1	140	1
11.	Sturman, D.	1	140	1

TABLE 13.2 Country-wise Analysis: Most Worked Nation in the Field of Wearable Technologies

SL. No.	Country	Documents	Citations	Total Link Strength
1.	United States	85	1,814	37
2.	United Kingdom	59	1,308	33
3.	China	12	406	6
4.	Netherlands	20	393	12
5.	Australia	18	359	14
6.	Italy	17	277	12
7.	Spain	10	273	10
8.	Germany	10	159	17
9.	Canada	14	157	7
10.	Finland	12	139	8

13.3.3 WEARABLE TECHNOLOGIES DEVELOPMENTS AND INTEGRATED BUSINESS MODEL

The market for wearable devices has expanded rapidly, allowing manufacturers to produce a wide range of devices to meet the needs and wants of consumers (Mardonova & Choi, 2018). In addition to providing users with comfortable, consistent, and hassle-free access, wearable products

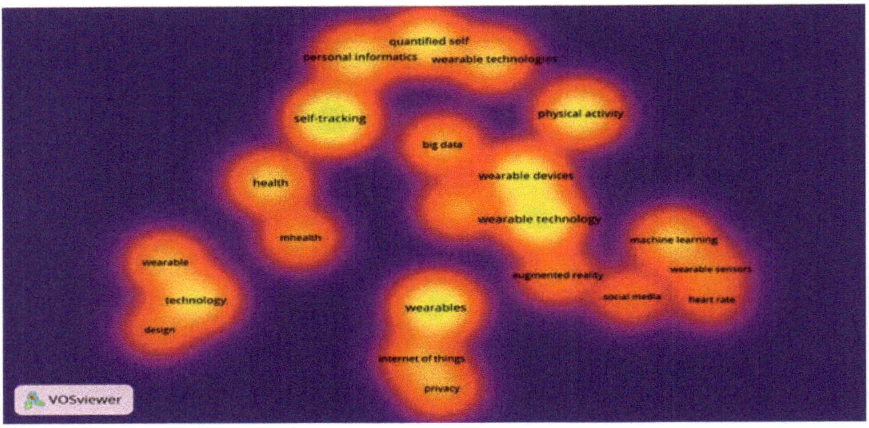

FIGURE 13.3 Keyword analysis of wearable technologies.

TABLE 13.3 Keyword Analysis of Wearable Technologies: Occurrence and Total Link Strength

SL. No.	Keyword	Occurrences	Total Link Strength
1.	Wearable technology	51	15
2.	Wearables	24	15
3.	Wearable technologies	15	9
4.	Physical activity	12	12
5.	Machine learning	12	8
6.	Wearable devices	10	13
7.	Self-tracking	9	18
8.	Wearable	8	7
9.	Internet of things	8	6
10.	Technology	7	10
11.	Quantified self	6	11
12.	Health	6	9
13.	Big data	6	7
14.	Augmented reality	6	5
15.	Wearable sensors	5	4
16.	Social media	5	4
17.	Privacy	5	6
18.	Personal informatics	5	10
19.	mHealth	5	6
20.	Heart rate	5	3
21.	Education	5	6
22.	Design	5	4

are designed to make users more aware of their surroundings by relaying information about their individual and environmental conditions and by layering computer-obtained images or sound files over the actual world (Billinghurst & Starner, 1999). Various industries (Mardonova & Choi, 2018) make use of wearable devices, such as the medical aid sector for data collection on patient health (Heintzman, 2016). Whether or not they are worn on the body, fashion industry gadgets can be found in a variety of aesthetically pleasing accessories (Cantanhede, Dias, Gammarano, & Arruda Filho, 2018).

13.3.3.1 BUSINESS CANVAS

It is also crucial to consider "the technology ecosystem in which it operates" (Dubs & Koschell, 2018), which includes key performance indicators for the business modeling of wearable technologies such as: important allies, significant operations, and value addition to the chain. The customer, considered as the king, is key to maintaining a healthy relationship with them in this business canvas (Hidefjall & Titkova, 2015). Segmentation describes the major domains and classifications where the pitch and research can be conducted. Cicek (2015) states that facilitators are the major resource, and channels/mediums are the roots of the supply chain mechanism. Moving towards the end, investment centers need to keep in mind R&D and product development as major areas of concern. Direct sales to end consumers, corporate sales, rental fees, consultancy services, data processing and overview, subscription fees, installation and commissioning services, and configuration and algorithm training will be the profit centers and outcomes of this canvas (Figure 13.4).

13.4 ISSUES, CONTROVERSIES, AND PROBLEMS

It is clear that wearable technology will play a pivotal role in the evolution of ICT infrastructure. Data procurement and computation, communications, password protection, confidentiality, equipment constraints, and consumer enactment are just some of the major problems that need to be addressed at this early stage. This study focuses on these issues and provides readers with a comprehensive overview of prospective options for overcoming the current challenges. The most significant hurdles in the production and sale of wearables are cost and a lack of manufacturing infrastructure.

<table>
<tr><th>Important Allies</th><th>Important Operations</th><th>Value Prepositions</th><th>Partnering with customers</th><th>Segmentation</th></tr>
<tr><td rowspan="3">Operating System Providers
Application Developers
Activity Tracking Devices
Analytical Software's
Camera Application
Developers
AI, IOT and AR
Chemical and Biological
Sensors developers
Analog Devices
Wearable devices
Manufactures</td><td>Health & fitness
Wellness Management
Information and appeal to fashion
Virtual Assistance</td><td rowspan="3">Building a stronger connection between people and their health
Providing contextual intelligence
Control and automation
Diverse styles and design aesthetics
Bio-sensing</td><td>Dedicated virtual assistance available round the clock
Provide fitness statistics
Extensive protection with the help of chemical and biological sensors
Supplied pet locations and heath related alerts</td><td rowspan="3">Healthcare
Consumer applications
Emergency Responder
Defence
Fitness & sports
Lifestyle
Animals</td></tr>
<tr><th>Facilitators</th><th>Channels</th></tr>
<tr><td>Easy and open interface for app developers
Core Team
Channel Partners
Funding Agencies
Customer Relationship Management</td><td>Purchase online or In store
Involving key partners to reach consumers
Awareness and promotion also through social media
Provide virtual after sales services</td></tr>
<tr><th colspan="2">Investment Centres</th><th colspan="3">Profit Centres</th></tr>
<tr><td colspan="2">Research and Innovation
Fixed Cost for Core Team
Product Development
Channel Expenses</td><td colspan="3">Direct Sales to end consumer, Corporate sales, Rental Fees, Consultancy Services, data processing and overview, Subscription Fees, Installation and commissioning services, Configuration, algorithm training</td></tr>
</table>

FIGURE 13.4 Business model canvas for wearable devices.

Poor merchandise effectiveness is not a deciding factor for today's consumers, especially among the initial majorities, as evidenced by the low rate at which new customers are retained. This has been a formidable obstacle to the acceptance of WT's products. Managers have suggested increasing spending on marketing initiatives, such as advertising and labeling, to overcome this problem.

Adopting wearables solely for the purpose of adopting wearables provides inadequate motivation to acquire and insufficient persuasion to overcome the lack of value-added services. Therefore, beyond intellectual curiosity, external factors such as practical encouragement, such as establishing connections in a unique context, situational highlights embedded in products, and peer interaction functionalities, can compensate for a lack of valuable offerings.

Interconnection, or the ability of different WT products to work together, is an important factor in reducing market fragmentation, increasing the value of WT to target consumers, and fostering innovation.

Third, reorienting entrepreneurial procedures: using user-generated data to counteract rigorous class divisions that lead to presumptuous models. Involving current and future users and other stakeholders in ideation exercises that can be used to create features and designs is highly beneficial. The sections on product design and development will offer advice on appropriate methodologies to implement this market-driven strategy.

13.5 FUTURE DIRECTIONS AND CONCLUSION

Technologies and gadgets had previously been in our hands in the form of cell phones, but they are now evolving to be smarter and more compact, and they have even made their way into our closets. Devices are becoming more diverse in terms of the attributes they offer, and they are also becoming more wearable. There is a high demand in this industry. These intelligent wearables can be used for a variety of purposes. What sets them apart from other bulky or heavy electronics is the fact that they are hands-free.

Microchip implantation is increasingly being utilized to replace keys and passwords as they evolve and become smarter. Near-field communication (NFC) or radio-frequency identification (RFID) chips are implanted on your fingertip (RFID). According to IDC, intelligent compacts will generate strong demand in the future. There are also in line, such as 'GOQII,' who are attempting to capitalize on this enormous potential. Whatever the future brings, one thing is certain: it will be fascinating.

KEYWORDS

- **bibliometrics analysis**
- **business model canvas**
- **integrated business model**
- **near-field communication**
- **radio-frequency identification**
- **wearable technology**

REFERENCES

Adams, R., (2018). Machine learning methods for activity detection in wearable sensor data streams. *Doctoral Dissertations*, 1318. https://doi.org/10.7275/0nsf-5682 https://scholarworks.umass.edu/dissertations_2/1318.

Billinghurst, M., & Starner, T., (1999). New ways to manage information. *Computer, 32*, 57–64. 10.1109/2.738305.

Chawla, N., (2020). AI, IOT and wearable technology for smart healthcare? A review. *Int J Green Energy*, *7*(1), 9–13.

Çiçek, M. E. S. U. T., (2015). Wearable technologies and its future applications. *International Journal of Electrical, Electronics and Data Communication*, *3*(4), 45–50.

Dehghani, M., & Dangelico, R., (2018). Smart wearable technologies: State of the art and evolution over time through patent analysis and clustering. *International Journal of Product Development, 22*, 293. 10.1504/IJPD.2018.091148.

Dubs, K., & Koschell, K., (2018). *Wearable Devices: A Technological Trend with Implications for Business Models – Master's Thesis in Business Administration Strategy and Management in International Organizations Advanced Level.* 30 credits Spring semester 2018 ISRN: LIU-IEI-FIL-A--18/02845--SE.

Dunne, L. E., (2004). *The Design of Wearable Technology: Addressing the Human-Device Interface Through Functional Apparel Design.*

Gao, Y., Li, H., & Luo, Y., (2015). An empirical study of wearable technology acceptance in healthcare. *Industrial Management & Data Systems.*

Godfrey, A., Hetherington, V., Shum, H., Bonato, P., Lovell, N. H., & Stuart, S., (2018). From A to Z: Wearable technology explained. *Maturitas, 113*, 40–47. Doi: 10.1016/j.maturitas.2018.04.012. Epub 2018 Apr 25. PMID: 29903647.

Heintzman, N. D., (2016). A digital ecosystem of diabetes data and technology: Services, systems, and tools enabled by wearables, sensors, and apps. *Journal of Diabetes Science and Technology*, *10*(1), 35–41. doi: 10.1177/1932296815622453.

Hidefjäll, P., & Titkova, D., (2015). Business model design for a wearable biofeedback system. In: *pHealth* (pp. 213–224). doi: 10.3233/978-1-61499-516-6-213.

Kurwa, M., Mohammed, A., & Liu, W., (2008). *Wearable Technology, Fashioning the Future.* Flextronics. Retrieved from: http://www.flextronics.com/Documents/Whitepapers/WTFashioningtheFuture-FINAL_AE.PDF.

Lazaroiu, G., Bratu, S., Serban, S., & Iordanescu, M., (2013). The semantics of communicative functions of smart interactive clothing. *Industria Textila., 64*, 260–265.

Mardonova, M., & Choi, Y., (2018). Review of wearable device technology, and its applications to the mining industry. *Energies*, *11*(3), 547. https://doi.org/10.3390/en11030547.

Moher, D., Shamseer, L., Clarke, M., et al., (2015). Preferred reporting items for systematic review and meta-analysis protocols (PRISMA-P) 2015 statement. *Syst Rev.,4*, 1. https://doi.org/10.1186/2046-4053-4-1.

Porter, M. E., & Heppelmann, J. E., (2014). How smart, connected products are transforming competition. *Harvard Business Review*, *92*(11), 64–88.

Profita, H. P., (2011). *Social Acceptability of Wearable Technology Use in Public: An Exploration of the Societal Perceptions of a Gesture-Based Mobile Textile Interface.*

Raj, D., & Ha-Brookshire, J., (2016). How do they create superpower'? An exploration of knowledge-creation processes and work environments in the wearable technology industry. *International Journal of Fashion Design, Technology and Education., 9*, 82–93. 10.1080/17543266.2015.1133720.

Rastogi, A., Daim, T., & Tan, J., (2008). Charting health information technology futures for healthcare services organizations. *International Journal of Healthcare Information Systems and Informatics (IJHISI)*, *3*(1), 1–23.

Schreurs, W., Hildebrandt, M., & Kindt, E., (2022). In: Tehrani, K., & Michael, A., (eds.), Wearable technology and wearable devices: everything. *Elgar Encycl. Law Data Sci.* p. 240.

Tehrani, K., & Andrew, M., (2014). *Wearable Technology and Wearable Devices: Everything You Need to Know.* Wearable Devices Magazine.

CHAPTER 14

Motivations, Key Success Factors, Functionality, and the Role of Social Entrepreneurship in Socio-Economic Sustainability: A Systematic Review of Existing Literature

ARUN MITTAL,[1] PRIYANK KUMAR SINGH,[2] and P. HORSLEY SOLOMON[3]

[1]*Birla Institute of Technology, Mesra, Ranchi, Off-Campus Noida, Uttar Pradesh, India*

[2]*Doon University, Dehradun, Uttarakhand, India*

[3]*Department of Electronics Science, SRM Arts and Science College, Kattankulathur, Chengalpattu, Tamil Nadu, India*

ABSTRACT

Social entrepreneurship (SE) encompasses the characteristics of both social organizations and business ventures. SE operates on the principle of socio-economic sustainability through a self-sustaining approach. However, further research is needed to identify the variables that contribute to the motivations, critical success factors, functionality, and role of SE. This systematic literature review examines 66 high-quality papers focused on SE, providing a detailed discussion of the aforementioned four parameters. Existing literature reveals that personal, social, and cultural factors drive social entrepreneurs to engage in such ventures. Moreover, SE requires

Sustainable Strategic Business Infrastructure Development and Contemporary Digital Practices in Industry 5.0. Indrajit Ghosal, Bikram Prasad, Nilanjan Ray, & Lalit Kumar Joshi (Eds.)

financial and sociocultural support to thrive and make significant contributions to employment, social upliftment, and overall societal development.

14.1 INTRODUCTION

Social entrepreneurship (SE) has gained significant popularity among industry and academia. The former conceptualizes social enterprises, while the latter studies and explores them. Academics are proposing and developing models driven by value creation in SE. Conventional thinking and the status quo worldwide have been challenged. Existing literature primarily focuses on social, ethical, and environmental concerns (Cortes & Lee, 2021). Interest in SE has been attracted in literature as a means to address social issues and enhance societies and communities. Social enterprises exist at the intersection of civil society, public policies, and the market. They take collective action to pursue social goals, with constraints on profit distribution and a participatory and open decision-making and ownership model (Molecke & Pinkse, 2017). The economic benefits of addressing the root causes of poverty through SE appear promising. Businesses created by social entrepreneurs may be for-profit or not, but their focus is almost always on using market-oriented activities to bring about systemic change that improves people's daily lives. The financial ecosystem presents a unique opportunity for social entrepreneurs (Kazmi et al., 2016). Engaging simultaneously or at higher levels across multiple domains can have harmful effects on bricolage. Furthermore, bricolage has three other adverse impacts: fruitless struggles, less cumulative development, and the inability to engage suppliers and customers (Senyard et al., 2014).

14.2 METHODOLOGY

The present chapter is a systematic literature review on various dimensions of SE, such as the motivations of SE, the functionality of SE, critical success factors of SE, and the role of SE in social and economic development. The studies have been obtained from renowned sources such as Taylor & Francis, Emerald, Inderscience, and other reputed publishers. The study contains 66 high-quality research papers, including empirical studies, case studies, and a systematic literature review. Only those papers were selected with the keywords "social entrepreneurship" or "social enterprise" in their title or abstract. Initially, 203 studies were shortlisted, and based on the criteria of the last 10-year time horizon, 66 studies were selected. The year-wise distribution of the studies has been presented in Table 14.1 and Figure 14.1.

TABLE 14.1 Year Wise Distribution of Studies

Year	Number of Studies	Percentage (%)
2022	10	15.2
2021	9	13.6
2020	7	10.6
2019	6	9.1
2018	7	10.6
2017	5	7.6
2016	4	6.1
2015	6	9.1
2014	10	15.2
2013	2	3.0
Total	66	100.0

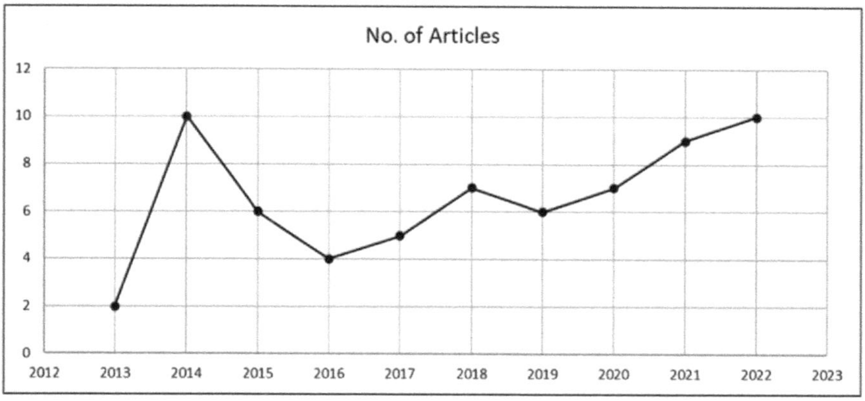

FIGURE 14.1 Year-wise distribution of studies.

It can be observed from Table 14.1 and Figure 14.1 that approximately 60% of the studies were conducted within the last five-year period, specifically from 2018 to 2022. The remaining 40% of the studies are from the preceding five-year period. This provides a comprehensive representation of the most recent studies in the field over the span of ten years.

Table 14.2 shows the sources of articles in this study. The majority of papers have been sourced from the Journal of Social Entrepreneurship (12), followed by the Journal of Business Research (4) and Social Enterprise Journal (4). The names of these journals have been mentioned as they are popular in the field of social enterprises and SE. Journals from which only one article was obtained have been categorized as "others." This category has contributed a total of 28 articles.

TABLE 14.2 Sources of Articles Under Study

SL No.	Name of the Journal	Number of Articles
1.	Journal of Social Entrepreneurship	12
2.	Journal of Business Research	4
3.	Social Enterprise Journal	4
4.	Journal of Business Venturing	3
5.	Entrepreneurship and Regional Development	3
6.	Journal of Asian Public Policy	2
7.	Business and Society	2
8.	International Journal of Social Entrepreneurship and Innovation	2
9.	Journal of Developmental Entrepreneurship	2
10.	Journal of Entrepreneurship in Emerging Economies	2
11.	New England Journal of Entrepreneurship	2
12.	Others	28
	Total	**66**

14.3 ANALYSIS AND DISCUSSION

14.3.1 MOTIVATIONS FOR SOCIAL ENTREPRENEURSHIP (SE)

There are a variety of motivations for social entrepreneurship (SE), including personal and social management of crises, community development, social change (Hamby, Pierce, & Brinberg, 2017), support during wars (Muhammad, Khan, Shah, & Ali, 2020), and exclusive government support. Galindo Martin et al. (2021) found that successful events and government failures in resolving complex issues have created opportunities for many social enterprises. Social entrepreneurs learn from both and contribute to the promotion of economic development and social reality. Their efforts include supporting disabled and homeless individuals, assisting during climate change conditions, and more. Issues may arise due to crises such as financial crises and global crises (Caldwell, Parker Harris & Renko, 2020). Motivations also include economic utility and social efficiency in emerging economies (Kabbaj, El Ouazzani Each Hadi, Elamrani, & Lemtaoui, 2016).

Life and a strong orientation towards high achievement, aimed at improving life and developing influence, are major drivers for social entrepreneurship (SE). Successful entrepreneurship in SE focuses on changing systems at both local and national levels. Success in SE can be seen when social entrepreneurs achieve their goals, are no longer needed, and become obsolete (Wanyoike & Maseno, 2021). Similarly, Chandra (2017) suggests that spirituality or religiosity provides a comprehensive framework for understanding why individuals show compassion and support others who may be less fortunate, engaging them in SE. Various religions encompass doctrines and values that promote a blend of commercial, communal, and social activities. Spirituality, faith, kindness, and commercial activities have been prevalent worldwide even before the concept of SE emerged. Past experiences and achievement orientation are two key motivational factors that drive the success of social entrepreneurs, enabling them to build capacity and bring about systematic transformation. This ultimately leads to the fulfillment of SE, where social entrepreneurs or enterprises become immaterial. The motivational elements of SE are closely tied to addressing societal issues. To promote SE, there needs to be a mindful awareness when starting an enterprise (Elliot, 2019). Understanding the impact of resistance or challenges on the interface of a mediator within a complex system provides insights into the relationship between entrepreneurial action and social change (Tan & Yoo, 2015).

Urban development and social entrepreneurship (SE) are closely associated with and support each other. The synchronization of SE policies and the growth of the urban sector is increasing. The association between SE and the growth of urban areas is an interesting research topic. Another area of growth in SE is the increasing number of older populations in different parts of the world and their contributions to development through SE channels. The age group and stage in the life cycle also serve as motivation for SE, as it now attracts young people and millennials (Caringal-Go & Hechanova, 2018).

Additionally, newly retired individuals can be effectively utilized as valuable connections for this channel (Angelidou & Mora, 2019).

Social entrepreneurship can be encouraged through empathy, which is associated with both social worth and self-efficacy. It has been observed that empathy serves as a precursor to both self-efficacy and social worth (Bacq & Alt, 2018).

Many studies have highlighted that social entrepreneurship (SE) is an outcome of social needs. Factors such as the magnitude of poverty, ethnic group identity, political landscape resolution, low education rates, and corruption serve as motivational factors for successful social enterprises (Rivera Santos et al., 2015). In studying the motivational model of "blended SE," it was found that social entrepreneurs engage in SE for personal achievement and to assist society (Germak & Robinson, 2014; Wettermark & Berglund, 2022). Newth & Woods (2014) introduced a novel aspect to the motivations of SE, stating that resistance positively influences the outcome of socially entrepreneurial action. SE studies have shown that bricolage is a prevalent exercise among social entrepreneurship organizations (SEOs), whether they operate in an intimidating context characterized by a lack of resources or in a more substantial manner. Most SEs combine internally generated income with external donations. The outcomes for organizational growth in such circumstances may differ from those achieved through bricolage out of necessity. When SEs have high availability of resources and autonomy, they tend to lean towards bricolage. In this process, they select activities of their choice and benefit from resources to renew the enterprise more efficiently. In such circumstances, enterprises practice an ideology called bricolage. It is driven by recognizing perceived benefit instead of necessity. Additionally, there is increased awareness and a social attitude towards the shift in enterprise and SE, which captures the media's attention and the public's imagination. This also raises the question of a parallel increase in the action of social entrepreneurs as an outcome or reason for a shift in attitude (Desa & Basu, 2013). In a different motivation, Geradts & Alt (2022) found that established organizations present unique opportunities and challenges to individuals engaging in entrepreneurial action to address social problems that remain unexplored. Furthermore, a study also found that sustainable development goals (SDGs), particularly basic amenities like drinking water, sanitation, and waste management, serve as motivation for SE (Goyal, Agrawal, & Sergi, 2020; Diaz-Sarachaga & Ariza-Montes, 2022). Social enterprises are alternative organizations that lead large-scale social change (Dey, Schneider, & Maier, 2016). In the Latin American context, it was found that socio-political activism increases experience, entrepreneurial passion, and social entrepreneurial intention among entrepreneurs in Latin America. Entrepreneurial passion plays a significant role in social entrepreneurial opportunity recognition (Darden, 2022). The studies under the various sub-themes of SE Motivation are given in Table 14.3.

TABLE 14.3 Studies focusing on various sub-themes of motivations for social entrepreneurship.

Sub-Themes	Studies
Personal factors: Awareness, achievement motivation, empathy, age, life cycle self-efficacy.	Wettermark & Berglund (2022); Darden (2022); Wanyoike & Maseno (2021); Muhammad, Khan, Shah, & Ali (2020); Angelidou & Mora (2019); Bacq & Alt (2018); Caringal-Go & Hechanova (2018); Germak & Robinson (2014) Tan & Yoo (2015); Desa & Basu (2013)
Socio-economic factors: Social issues, social change	Wanyoike & Maseno (2021); Muhammad, Khan, Shah, & Ali (2020); Bacq & Alt (2018); Hamby, Pierce, & Brinberg (2017); Chandra (2017); Kabbaj, El Ouazzani Ech Hadi, Elamrani, & Lemtaoui (2016); Rivera Santos et al. (2015); Germak & Robinson (2014); Desa & Basu (2013)
Social resistance: Challenges as motivation and support	Dey, Schneider, & Maier (2016); Tan & Yoo (2015); Newth & Woods (2014)
Sustainable development goals: Basic amenities	Diaz-Sarachaga & Ariza-Montes (2022); Goyal, Agrawal, & Sergi (2021)
Political issues: Government success failure and crisis, wars	Galindo Martin et al. (2021); Caldwell, Parker Harris & Renko (2020); Rivera Santos et al. (2015)
System transformation	Wanyoike & Maseno (2021); Elliot (2019)

14.3.2 CRITICAL SUCCESS FACTORS FOR SOCIAL ENTREPRENEURSHIP (SE)

Regular growth and innovation are the most important prerequisites for the success of social enterprises. For the development of social entrepreneurs, a smooth and well-functioning institutional environment is crucial (Ahmad & Bajwa, 2021). In another study, the key success factors of SE were identified as capabilities for sustainable societal transformation and mobilizing innovative ideas (Dionisio, 2019). Smith, McVilly, McGillivray, & Chan (2018) found that social enterprises with consistent and reliable missions, as well as entrepreneurial competence that aligns with the trading environment, are more likely to receive donations from contributors. SE and bricolage share important characteristics that bring them closer together. However, bricolage in the context of SE has been under-studied and received less attention from scholars. Langevang & Namatovu (2019) examined the role of SEs after the war and discovered that social entrepreneurs engaged in various bricolage activities to establish social enterprises that mitigate the effects of war. SE business, at the same time, was also found to be "business as usual." Hence, it is observed that SE can also work efficiently and contribute during the nonexistence of extraordinary situations. Trust is a major component of SE. Bricolage is not just restrained to circumstances of scarcity but has the potential to positively influence the enterprise's growth (Bojana et al., 2018). It is sometimes even misunderstood with closer constructs like improvisation and effectuation. Many enterprises and groups of people aim to tackle social issues and, at the same time, create social values. There are three subsets of activities in which the process of entrepreneurship can be clustered: opportunity definition, resource leveraging, and organization building (Janssen, Fayolle, & Wuilaume, 2018). Nowadays, SE is supported as social intrapreneurship by business organizations (Geradts & Alt, 2022).

Entrepreneurs usually work in a context with limited resources. An environment with constrained resources may stem from various dimensions, including internal resources (such as a small number of employees and outdated skills), customers (lack of financial resources), and the business environment (fewer partners). The capabilities of bricolage can explain why some enterprises become successful compared to others when they face a scarcity of resources (Well et al., 2017). The role played by SE in resolving social and economic issues is noteworthy. Numerous research studies have been conducted on SESED, observing the role of SE in addressing social and economic issues in different nations around the world. Researchers argue that SE plays an essential role in resolving critical problems of society, such as economic inequalities, poverty, and resource wastage, in the modern economy. The studies under various subthemes of critical success factors for SE are presented in Table 14.4.

14.3.3 SOCIAL ENTERPRISE FUNCTIONALITY

Social enterprises generally operate similarly to other organizations; however, their functionality differs in some aspects. Their focus shifts according to social needs. In the context of Bangladesh, it was found that during the COVID-19 pandemic, social enterprises were primarily engaged in addressing food, healthcare, employment, and education (Huda & Maliha, 2021; Gigauri et al., 2022). Zhao (2020) explored how social enterprises can manage interpersonal relationships and goal achievement, emphasizing the importance of effective social skills in relational communication to explain the varying outcomes of social endeavors. Chandra & Paras (2020) examined how social entrepreneurs exhibit tremendous effectuation when dealing with a superstorm, gradually restoring it with intended effectuation over time. Social organizations demonstrated resilience during the COVID-19 pandemic and other crises. Richard (2020) discovered that the more diverse the capabilities possessed by social entrepreneurs, the greater the likelihood of achieving their targets. Wullman & Hudon (2015) examined the transition from social constructionists to social bricoleurs, which occurs with sufficient human and financial resources, ambition, determination, and significant accomplishments. A lack of financing hinders this transition.

TABLE 14.4 Studies focusing on various sub-themes of critical success factors for SEs.

Sub-Themes	Studies
Regular growth, innovation, smooth functioning environment, and institutional support.	Ahmad & Bajwa (2021); Dionisio (2019)
Capabilities of bricolage.	Witell et al. (2017); Janssen, Fayolle, Wuilaume (2018). Langevang & Namatovu (2019); Bojica et al. (2018); Wullman & Hudon (2015)
Opportunity definition, resource leveraging, and building of organization.	Janssen, Fayolle, & Wuilaume (2018)
Consistent and reliable mission.	McVilly, McGillivray & Chan (2018)
Social intrapreneurship.	Geradts & Alt (2022)
Capabilities of social transformation.	Dionisio (2019)

In conclusion, the results indicate that social constructionists typically opt for a hybrid or mixed business model that combines profit and non-profit elements. This hybrid structure allows Mexican social enterprises to rely less on contributions or donations and instead focus on their vision as social enterprises. Vealey & Gerding (2016) explored how crowdfunding has become a new way of managing finance for social enterprises. To ensure effective functionality, social enterprises should prioritize social innovation processes (Mueller et al., 2015; Jarrar, 2022). Parente, Lopes, & Marcos (2014) found that the status of social enterprises is granted to associations that employ strategic management, with the outcome being performance rather than profit. Littlewood & Holt (2018) recommended that social entrepreneurs leverage their social skills to cultivate interpersonal relationships with actors, thereby enhancing reputation and fostering a positive image among members of the social enterprise. The studies related to the various sub-themes of social enterprise functionality are presented in Table 14.5.

TABLE 14.5 Studies Focusing on Various sub-themes of the Functionality of SEs

Sub-Themes	Studies
Need specific and flexible functioning	Huda & Maliha (2021); Gigauri et al. (2022); Chandra& Paras (2020)
Interpersonal relationships, social skills, and goal accomplishment	Zhao (2020); Littlewood & Holt (2018)
Diversity to the target	Richard (2020)
Hybrid or mixed business model among profit & non-profit.	Wullman & Hudon (2015)
Crowdfunding	Vealey & Gerding (2016)
Social innovation processes	Mueller et al. (2015); Jarrar (2022)
Professional and career development	Parente, Lopes, & Marcos (2014)

14.3.4 THE ROLE OF SOCIAL ENTREPRENEURSHIP (SE)

SE can help achieve sustainable development, financial independence, improve health and well-being, and advocate for human rights (Bansal et al., 2019; Roy & Roy, 2020). In the context of "Social Enterprise in Socio-Economic Development" or SESED, it has been found that social enterprises contribute to social influence, resource mobilization, and bring about sustainable social changes, ultimately leading to socio-economic development, including urban development (El Ebrashi, 2013; Angelidou & Mora, 2019). The building block of SE is the development of systematic transformation. It is believed that organizations have a long way to go before management and organizational practices reach stable levels to ensure financial sustainability. Additionally, it is essential to note that practices of control and accountability are completely lacking in the profiles of national SE (Parente, Lopes, & Marcos, 2014). Sustainability is a benchmark of an outcome that reflects the achievement of social entrepreneurial associations. Social entrepreneurship is a constant and continuous process, and the initiatives that remain relevant and current enjoy long-term sustainability. It is observed that social entrepreneurs must possess capabilities such as resource bricolage, which is the skill to maximize results from available resources in order to achieve long-term goals (Zhang & Swanson, 2014). For the welfare of the labor market, there is an argument that the diffusion of technology in driving entrepreneurship is important to increase employment capacity. Social entrepreneurship and technology diffusion have a significant influence on future employment (Osabohien, Worgwu, & Al-Faryan, 2022). Economic and social entrepreneurs share similar emphasis on opportunity and vision, as well as the ability to persuade and inspire others to help turn a vision into reality (Osabohien et al., 2022). Social entrepreneurship contributes to resolving social issues and makes efforts towards social and economic development. It encourages the building of communities and the cultivation of culture, along with creative methods of addressing socio-economic issues that ultimately lead to economic growth. It also addresses issues such as literacy, poverty, labor or human rights, public health, water or energy-related issues, the environment, social inequality, and unemployment (Nega & Schneider, 2014; Pathak & Muralidharan, 2018).

In Jordan, social entrepreneurship (SE) has also contributed to women's inclusion and assisting refugees (Jarrar, 2022). As a sub-domain of entrepreneurship, SE supports a bidirectional association with social and economic development. Entrepreneurship has been encouraged by the economic development of a nation, both commercially and socially. Scholars have identified various determinants of SE, including national culture, entrepreneurial ecosystem, innovation, and entrepreneurial spirit (Kivarina et al., 2020). Social enterprises are considered suitable solutions in nations with significant inequalities and corrupt governments, spreading awareness among people (Dufays & Huybrechts, 2014). Scholars have also argued that market-based solutions can be implemented for poverty through socially accountable businesses with SE programs to create social values (Zaefarian et al., 2015).

SE contributes to economic development by addressing issues such as poverty alleviation, education, healthcare, civil engagement, environment, disability, aging, and more. Extensive research has been conducted on economic development, particularly in the area of poverty alleviation. SE plays a crucial role in supporting underprivileged populations, including the disabled, poor, women, and uneducated, who face challenges in accessing traditional job opportunities (Roy, Donaldson, Baker, & Kerr, 2014). Younis et al. (2020) found that social support acts as a mediator, reinforcing the impact of social entrepreneurial self-efficacy on SE intentions. Social entrepreneurs are dedicated to creating positive change in society, emphasizing sustainability, value practices, innovation, and the ability to build social networks and generate viable financial income. Social enterprises have contributed to social impact by expanding services, fostering new collaborations between organizations, and reaching more beneficiaries (Loukopoulos & Papadimitriou, 2022). Table 14.6 provides an overview of studies conducted on various sub-themes related to the role of SE.

TABLE 14.6 Studies focusing on various sub-themes of the role of SEs.

Sub-Themes	Studies
Socio-economic impact: Resolving social issues, social impact, and values, social, and economic development.	Loukopoulos & Papadimitriou (2022); Jarrar (2022); Kivarina et al. (2020); Roy & Roy (2020); Younis et al. (2020); Angelidou & Mora (2019); Pathak & Muralidharan (2018); Zaefarian et al. (2015); Nega & Schneider (2014); Roy, Donaldson, Baker & Kerr (2014); Dufay's & Huybrechts (2014); El Ebrashi (2013)
Employment generation.	Osabohien et al. (2022); Worgwu & Al-Faryan (2022); Osabohien, Roy & Roy (2020); Pathak & Muralidharan, 2018
Innovations and capacity building: New business opportunities, systematic transformation.	Osabohien et al. (2022); Younis et al. (2020); Kivarina et al. (2020); Zaefarian et al. (2015); Parente, Lopes & Marcos (2014)
Sustainability: Sustainable finance, sustainable development.	Diaz-Sarachaga & Ariza-Montes (2022); Bansal et al. (2019); Parente, Lopes & Marcos (2014); Zhang & Swanson (2014)

14.4 FINDINGS OF THE STUDY

The objective of this study was to connect social entrepreneurship (SE) with socio-economic sustainability. This chapter mainly discusses four aspects of SE–'motivations for SE,' 'key success factors of SE,' 'SE functionality,' and 'the role of SE in socio-economic sustainability.' Motivation for SE depends on personal factors such as awareness, achievement motivation, empathy, age, life cycle, self-efficacy, social issues, social change, social resistance, challenges to sustainable development goals (SDGs), crises, wars, and system transformation. Key success factors of SE include regular growth, innovation, a smooth functioning environment, institutional support, bricolage capabilities, opportunity definition, resource leveraging and building of organization, consistent and reliable mission, social intrapreneurship, and capabilities of social transformation. The third aspect of SE, i.e., SE functionality, requires flexibility, interpersonal relationships, social skills, and goal accomplishment. It also involves diversity in the target hybrid or mixed business model among profit and non-profit enterprises, crowdfunding, social innovation processes, and professional and career development. Lastly, in the existing literature, the role of SE has been identified as resolving social issues, promoting social impact and values, contributing to social and economic development, generating employment, fostering innovations, and facilitating capacity building. It also presents new business opportunities, enables systematic transformation, promotes sustainable finance, and contributes to sustainable development.

14.5 DISCUSSION AND CONCLUSION

Social entrepreneurs play a vital role in economic development in the modern world (Galindo Martin et al., 2021). Many empirical studies have found a significant impact of SE on poverty alleviation, socioeconomic development, and sustainability (Rivera Santos et al., 2015). SE receives support from various other disciplines such as management, social work, philosophy, law, public policy, social policy, sustainability, sociology, and political science (Caldwell, Parker, & Renko, 2020). Nowadays, young people and millennials have shown interest in SE as key participants. Additionally, a country's culture, socio-economic development, and governance institutions also support SE (Puumalainen, Sjogren, Syrja, & Barraket, 2015).

A case study on Kaaita established that an innovative business model is one important prerequisite of successful SE (Pearse & Peterlin, 2019). Top managers need to recognize social enterprise creation as feasible or desirable. The intentions of SE would be few or absent (Tan & Yoo, 2015). Like any other profit or non-profit venture, SE has challenges, too, and profit and financial sustainability top the list. To overcome the challenge, effective management of operating expenses and promotion of enterprise, particularly infrastructure, is required (Boluk & Mottiar, 2014). Entrepreneurial awareness and encouragement, organization, institution, society, and market in which entrepreneurship is surrounded. One of the relatable issues, particularly in SE, is resistance, as enterprises possibly face resistance across numerous fronts while achieving their social transformations and efforts in launching and achieving justifiable economic processes (Newth & Woods, 2014). The planned behavior theory recommends that entrepreneurial intentions influence future actions (Baierl, Grichnik, Sporrle, & Welpe, 2014). The multidimensional structure suggests that social entrepreneurs maintain a blended motivation that includes elements of achievement orientation, self-actualization, and a desire to contribute to society (Germak & Robinson, 2014). Political philosophy helps to acknowledge the fundamental social aspect of "social entrepreneurship."

Similarly, normative awareness assists in determining the social, economic, and cultural impact of SE (Swaim & Roebuck, 2021).

Furthermore, Swaim & Roebuck (2021) recommended investigating the ethical framework of SEs through a three-dimensional model that includes "ethical empowering leadership," "perceived ethical organizational support," and "empowered employee ethical voice SE." This model has covered a wide variety of issues. However, there are still many other issues that need to be addressed (Pathak & Muralidharan, 2018). These issues may be related to people's physical and mental health, as this topic has gained significant attention in the post-COVID-19 scenario. Non-profit organizations often face inadequate financial resources in their respective sectors (Dufays & Huybrechts, 2014). SE portrays those who run the social enterprise as "heroic, creating economic and social value," while the people they serve are often poor and in need (Wettermark & Berglund, 2022). SE primarily focuses on the originators and their team members, who are typically mature individuals. Researchers often examine the social enterprise founders and their employees (Caringal-Go & Hechanova, 2018).

This study recommends a step-by-step and empirical investigation of SE dimensions. The study has explored the role of SE in socio-economic sustainability and has connected SE with its various dimensions. Diversity and trust have been found to be the most important factors contributing to the successful implementation of SE (Ayob, 2017). Social enterprises must create socio-economic value by addressing complex social problems and balancing the roles of the public and private sectors (Dionisio, 2019). In conclusion, SE includes ethical, personal, emotional, psychological, economic, and political perspectives. This makes the role of SE complex. However, gradually, the stakeholders must improve, all while keeping a good balance inter-se.

KEYWORDS

- **non-profit organizations**
- **social enterprise**
- **social entrepreneurship**
- **social entrepreneurship motivations**
- **socio-economic development**
- **socioeconomic sustainability**
- **sustainability**

REFERENCES

Ahmad, S., & Bajwa, I. A., (2021). The role of social entrepreneurship in socio-economic development: A meta-analysis of the nascent field. *Journal of Entrepreneurship in Emerging Economies.*doi 10.1108/JEEE-04-2021-0165, 2053-4604.

Angelidou, M., & Mora, L., (2019). Developing synergies between social entrepreneurship and urban planning evidence from six European cities. *disP, 55*(4), 28–45.

Ayob, A. H., (2017). Diversity, trust, and social entrepreneurship. *Journal of Social Entrepreneurship,* 1–12. doi: 10.1080/19420676.2017.1399433.

Bacq, S., & Alt, A., (2018). Feeling capable and valued: A prosocial perspective on the link between empathy and social entrepreneurial intentions. *Journal of Business Venturing, 33*(3), 333–350. doi: 10.1016/j.jbusvent.2018.01.004.

Baierl, R., Grichnik, D., Sporrle, M., & Welpe, I. M., (2014). Antecedents of social entrepreneurial intentions: The role of an individual's general social appraisal. *Journal of Social Entrepreneurship, 5*(2), 123145.

Bansal, S., Garg, I., & Sharma, G. D., (2019). Social entrepreneurship as a path for social change and driver of sustainable development: A systematic review and research agenda. *Sustainability, 11*(4), 1091. https://doi.org/10.3390/su11041091.

Bojica, A. M., Jimenez, J. M. R., Nava, J. A. R., & Fuentes-Fuentes, M. M., (2018). Bricolage and growth in social entrepreneurship organizations. *Entrepreneurship & Regional Development, 30*(3), 362–389.

Boluk, K., & Mottiar, Z., (2014). Motivations of social entrepreneurs. *Social Enterprise Journal, 10*(1), 53–68.

Caldwell, K., Parker, H. S., & Renko, M., (2020). Inclusive management for social entrepreneurs with intellectual disabilities: "How they act. *Journal of Applied Research in IntellectualDisabilities, 33*(2), 204–218. https://doi.org/10.1111/jar.12662.

Caringal-Go, J. F., & Hechanova, M. R. M., (2018). Motivational needs and intent to stay of social enterprise workers. *Journal of Social Entrepreneurship, 9*(3), 200–214. https://doi.org/10.1080/ 19420676.2018.1468352.

Chandra, Y., & Kerlin, J. A., (2021). Social entrepreneurship in context: Pathways for new contributions in the field. *Journal of Asian Public Policy, 14*(2), 135–151. doi: 10.1080/17516234.2020.1845472.

Chandra, Y., & Paras, A., (2020). Social entrepreneurship in the context of disaster recovery: Organizing for public value creation. *Public Management Review*, 1–22. https://doi.org/10.1080/ 14719037.2020.1775282.

Chandra, Y., (2017). Social entrepreneurship as emancipatory work. *Journal of Business Venturing, 32*(6), 657–673. https://doi.org/10.1016/j.jbusvent.2017.08.004.

Cortes, A. F., & Lee, Y., (2021). Social entrepreneurship in SMEs: A note on three essential questions. *New England Journal of Entrepreneurship, 24*(2), 62–78. doi: 10.1108/ neje-03-2021-0014.

Darden, B. B., (2022). Political roots leading to opportunity recognition among Latin American social entrepreneurs. *Journal of Entrepreneurship and Public Policy, 11*(2, 3), 93–111. doi: 10.1108/jepp-01-2022-0013.

Desa, G., & Basu, S., (2013). "Optimization or bricolage? Overcoming resource constraints in global social entrepreneurship. *Strategic Entrepreneurship Journal,7*(1), 26–49.

Dey, P., Schneider, H., & Maier, F., (2016). Intermediary organizations and the hegemonization of social entrepreneurship: Fantasmatic articulations, constitutive quiescences, and moments of indeterminacy. *Organization Studies, 37*(10), 1451–1472. doi: 10.1177/ 0170840616634133.

Diaz-Sarachaga, J. M., & Ariza-Montes, A., (2022). The role of social entrepreneurship in the attainment of the sustainable development goals. *Journal of Business Research, 152*, 242–250. doi: 10.1016/j.jbusres.2022.07.061.

Dionisio, M., (2019). The evolution of social entrepreneurship research: A bibliometric analysis. *Social Enterprise Journal, 15*(1), 22–45.

Dufays, F., & Huybrechts, B., (2014). Connecting the dots for social value: A review on social networks and social entrepreneurship. *Journal of Social Entrepreneurship, 5*(2), 214237.

El Ebrashi, R., (2013). Social entrepreneurship theory and sustainable social impact. *SocialResponsibility Journal, 9*(2), 188–209.

Elliott, R. M., (2019). Social entrepreneurship as catalyst to break the poverty trap: An analysis of the motivational factors in South Africa. *Acta Commercii, 19*(2), 652.

Galindo-Martín, M. Á., Castaño-Martínez, M. S., & Méndez-Picazo, M. T., (2021). The role of entrepreneurship in different economic phases. *Journal of Business Research, 122*, 171–179.

Geradts, T. H., & Alt, E., (2022). Social entrepreneurial action in established organizations: Developing the concept of social intrapreneurship. *Journal of Business Research, 151*, 197–206. doi: 10.1016/j.jbusres.2022.06.047.

Germak, A. J., & Robinson, J. A., (2014). Exploring the motivation of nascent social entrepreneurs. *Journal of Social Entrepreneurship,5*(1), 5–21. doi: 10.1080/19420676.2013. 820781.

Gigauri, I., Panait, M., Apostu, S. A., & Raimi, L., (2022). The essence of social entrepreneurship through a Georgian lens: Social entrepreneurs' perspectives. *Administrative Sciences, 12*(3), 75. doi: 10.3390/admsci12030075.

Goyal, S., Agrawal, A., & Sergi, B. S., (2021). Social entrepreneurship for scalable solutions addressing sustainable development goals (SDGs) at BOP in India. *Qualitative Research inOrganizations and Management: An International Journal, 16*(3, 4), 509–529. doi: 10.1108/qrom-07-2020-1992.

Hamby, A., Pierce, M., & Brinberg, D., (2017). Solving complex problems: Enduring solutions through social entrepreneurship, community action, and social marketing. *Journal of Macromarketing, 37*(4), 369–380. doi: 10.1177/0276146716663797.

Janssen, F., Fayolle, A., & Wuilaume, A., (2018). Researching bricolage in social entrepreneurship. *Entrepreneurship & Regional Development, 30*(3), 450–470.

Jarrar, A., (2022). Entrepreneurship in Jordan: The ecosystem of the social entrepreneurship support organizations (SESOS). *Journal of Innovation and Entrepreneurship, 11*(1). doi: 10.1186/s13731-022-00200-z.

Kabbaj, M., El Ouazzani, E. H., K., Elamrani, J., & Lemtaoui, M., (2016). A study of the social entrepreneurship ecosystem: The case of Morocco. *Journal of Developmental Entrepreneurship, 21*(4), 1650021. doi: 10.1142/s1084946716500217.

Kazmi, S. S., Hashim, M., Kee, D. M., & Khan, F. U., (2016). Social entrepreneurship and its impact on economy: In perspective of Pakistan. *International Journal of Academic Research inAccounting, Finance and Management Sciences* (Vol. 6, pp. 161–166).

Kivarina, M. V., Kulikov, S. P., Zagorodnikov, K. A., Zhiltsov, V. A., & Khan, N. V., (2020). Economic growth humanization: Responsibility of business and social entrepreneurship. *Quality-Access to Success, 21*(176), 16–21.

Langevang, T., & Namatovu, R., (2019). Social bricolage in the aftermath of war. *Entrepreneurship & Regional Development, 31*(9, 10), 785–805. https://doi.org/10.1080/08985626.2019.1595743.

Littlewood, D., & Holt, D., (2018). Social entrepreneurship in South Africa: Exploring the influence of environment. *Business and Society, 57*(3), 525–561. https://doi.org/10.1177/0007650315613293.

Loukopoulos, A., & Papadimitriou, D., (2022). Organizational growth strategies for Greek social enterprises' social impact during the COVID-19 pandemic. *Social Enterprise Journal, 18*(4), 541–562. doi: 10.1108/sej-10-2021-0084.

Maliha, S. R., & Huda, S. M. S., (2021). "Rise of social entrepreneurship during COVID-19 pandemic. *International Journal of Social Entrepreneurship and Innovation, 6*(1), 63. Available at: https://doi.org/10.1504/ijsei.2021.10042995.

Molecke, G., & Pinkse, J., (2017). Accountability for social impact: A bricolage perspective on impact measurement in social enterprises. *Journal of Business Venturing, 32*(5), 550–568. doi: 10.1016/j.jbusvent.2017.05.003.

Mueller, S., D'Intino, R. S., Walske, J., Ehrenhard, M. L., Newbert, S. L., Robinson, J. A., & Senjem, J. C., (2015). What's holding back social entrepreneurship? Removing the impediments to theoretical advancement. *Journal of Social Entrepreneurship, 6*(3), 245–256. doi: 10.1080/19420676.2014.954259.

Muhammad, A., Khan, J. Z., Shah, S. I., & Ali, M., (2020). Exploring challenges and opportunities of the new social entrepreneurs: The case of indigenous musicpreneurs in KP Pakistan. *Journalof Entrepreneurship in Emerging Economies, 13*(5), 904–921. doi: 10.1108/jeee-03-2020-0063.

Nega, B., & Schneider, G., (2014). Social entrepreneurship, microfinance, and economic development in Africa. *Journal of Economic Issues, 48*(2), 367–376.

Newth, J., & Woods, C., (2014). Resistance to social entrepreneurship: How context shapes innovation. *Journal of Social Entrepreneurship, 5*(2), 192–213. doi: 10.1080/19420676.2014.889739.

Osabohien, R., Worgwu, H., & Al-Faryan, M. A. S., (2022). Social entrepreneurship, technology diffusion, and future employment in Nigeria. *Social Enterprise Journal*. doi: 10.1108/SEJ-03-2022-0032.

Osabohien, R., Worgwu, H., Adediran, O., & Soomro, J. A., (2022). Social entrepreneurship and future employment in Nigeria. *International Social Science Journal*. doi: 10.1111/issj.12360.

Parente, C., Lopes, A., & Marcos, V., (2014). Social entrepreneurship profiles: Lessons from organizational and management dynamics. *Journal of Social Entrepreneurship, 5*(1), 22–41. doi: 10.1080/19420676.2013.820782.

Pathak, S., & Muralidharan, E., (2018). Economic inequality and social entrepreneurship. *Business and Society, 57*(6), 1150–1190.

Pearse, N. J., & Peterlin, J., (2019). Artistic creative social entrepreneurs and business model innovation. *Journal of Research in Marketing and Entrepreneurship, 21*(2), 149–162. doi: 10.1108/jrme-07-2018-0036.

Puumalainen, K., Sjögrén, H., Syrjä, P., & Barraket, J., (2015). Comparing social entrepreneurship across nations: An exploratory study of institutional effects. *Canadian Journal of Administrative Sciences / Revue Canadienne Des Sciences De L'Administration, 32*(4), 276–287. doi: 10.1002/cjas.1356.

Ranville, A., & Barros, M., (2021). Towards normative theories of social entrepreneurship. A review of the top publications of the field. *Journal of Business Ethics, 180*(2), 407–438. doi: 10.1007/s10551-021-04867-4.

Richard, J. A., (2020). Modelling social entrepreneurship: Consideration of the reacting forces. *Journal of Social Entrepreneurship*, 1–18. doi: 10.1080/19420676.2020.1718744.

Rivera-Santos, M., Holt, D., Littlewood, D., & Kolk, A., (2015). Social entrepreneurship in sub-Saharan Africa. *The Academy of Management Perspective, 29*(1), 72–91.

Roy, M. J., Donaldson, C., Baker, R., & Kerr, S., (2014). The potential of social enterprise to enhance health and well-being: A model and systematic review. *Social Science & Medicine, 123*, 182–193. https://doi.org/10.1016/j.socscimed.2014.07.031.

Roy, M., & Roy, A., (2020). The evolution of social entrepreneurial initiatives by women in India: A longitudinal assessment across sectors. *International Journal of Social Entrepreneurship and Innovation, 5*(4), 277. doi: 10.1504/ijsei.2020.10038172.

Senyard, J., Baker, T., Steffens, P., & Davidsson, P., (2014). Bricolage as a path to innovativeness for resource-constrained new firms. *Journal of Product Innovation Management, 31*(2), 211–230.

Smith, P., McVilly, K. R., McGillivray, J., & Chan, J., (2018). Developing open employment outcomes for people with an intellectual disability utilizing a social enterprise framework. *Journal of Vocational Rehabilitation, 48*(1), 59–77. https://doi.org/10.3233/JVR-170916.

Swaim, J. A., & Roebuck, D., (2021). The role of empowered employee ethical voice in supporting social entrepreneurship. *Journal of Developmental Entrepreneurship, 26*(4). doi: 10.1142/s108494672150028x.

Tan, W., & Yoo, S., (2015). Social entrepreneurship intentions of non-profit organizations. *Journal of Social Entrepreneurship, 6*(1), 103–125. doi: 10.1080/19420676.2014.954260.

Vealey, K. P., & Gerding, J. M., (2016). Rhetorical work in crowd-based entrepreneurship: Lessons learned from teaching crowdfunding as an emerging site of professional and technical communication. *IEEE Transactions on Professional Communication, 59*(4), 407–427. doi: 10.1109/tpc.2016.2614742.

Wanyoike, C. N., & Maseno, M., (2021). Exploring the motivation of social entrepreneurs in creating successful social enterprises in East Africa. *New England Journal of Entrepreneurship, 24*(2), 79–104. doi: 10.1108/neje-07-2020-0028.

Wettermark, A., & Berglund, K., (2022). Mutuality between selves and others in social entrepreneurship: Not a mission impossible? *Scandinavian Journal of Management, 38*(3), 101219. doi: 10.1016/j.scaman.2022.101219.

Witell, L., Gebauer, H., Jaakkola, E., Hammedi, W., Patricio, L., & Perks, H., (2017). A bricolage perspective on service innovation. *Journal of Business Research, 79*, 290–298.

Wulleman, M., & Hudon, M., (2015). Models of social entrepreneurship: Empirical evidence from Mexico. *Journal of Social Entrepreneurship.* doi: 10.1080/19420676.2015.1057207.

Younis, A., Xiaobao, P., Nadeem, M. A., Kanwal, S., Pitafi, A. H., Qiong, G., & Yuzhen, D., (2020). Impact of positivity and empathy on social entrepreneurial intention: The moderating role of perceived social support. *Journal of Public Affairs.* https://doi.org/10.1002/pa.2124.

Zaefarian, R., Tasavori, M., & Ghauri, P. N., (2015). A corporate social entrepreneurship approach to market-based poverty reduction. *Emerging Markets Finance and Trade, 51*(2), 320–334.

Zhang, D. D., & Swanson, L. A., (2014). Linking social entrepreneurship and sustainability. *Journal of Social Entrepreneurship, 5*(2), 175–191.

Zhao, M., (2020). Social entrepreneurship for systemic change: The case of Southeast and South Asian countries. *Journal of Asian Public Policy.* https://doi.org/10.1080/17516234.2020.1796325.

CHAPTER 15

Healthcare Industry 4.0 to 5.0: A Psychological Study

BINDA KUMARI

Department of Psychology, K.O.C., Ratu Ranchi (Ranchi University), Ranchi, Jharkhand, India

ABSTRACT

Present research indicates that Industry 4.0 to 5.0 focuses on the use of a wide range of modern technologies in human healthcare. These technologies include digitization, response data (ergonomics), artificial intelligence (AI), human psychology, large data machine learning, and the internet of things (IoT). Currently, the human mind is ready to move towards Industry 5.0, and we need to review certain aspects of our mind or nature to further consider the industrial area. Healthcare Industry 4.0 is undergoing a paradigm shift thanks to Industry 5.0, which provides better user comfort and proactive intervention in early detection and treatment of various diseases. We have identified new modern trends in research and summarized features to guide and prepare the sector for an Industry 5.0 healthcare.

15.1 INTRODUCTION

In society, our mindset is that the new project is very useful for human development. Therefore, we are very excited to use this new product. Humans believe that our way of living and interacting in the digital world

Sustainable Strategic Business Infrastructure Development and Contemporary Digital Practices in Industry 5.0. Indrajit Ghosal, Bikram Prasad, Nilanjan Ray, & Lalit Kumar Joshi (Eds.)

is essential for the growth of our lives. Different social workers differ only in their abilities. Ultimately, people indicate that Healthcare Industry 5.0 is better than 4.0, which motivates workers to work with a fresh mind. When compared to other fields of scientific inquiry, the research in Healthcare Industry 4.0 is paving the way for 5.0.

15.2 MEANING OF INDUSTRY 4.0

Industry 4.0 is the digitization of knowledge in manufacturing, similar to other industries, and the creation of valuable processors. It is synonymous with the 4th industrial revolution and refers to manufacturing industries and control organizations in Industry 4.0. It enables remote and real-time monitoring with industrial and Internet of Things (IoT) technologies, in accordance with the principles of Industry 4.0.

The concept of Industry 4.0 has originated from the German manufacturing industry's efforts to be recognized worldwide. It also emphasizes the importance of end-to-end networking, from digital supply chains to all manufacturing processes, regardless of the number of intermediary steps or the end user.

There are four types of Industrial Revolution in Industry 4.0:

1. **1st Industrial Revolution:** The use of the 1st industrial revolution's powerful water and steam machinery led to a shift from manual production to machine-based production. Its effects were felt in various sectors, particularly in textiles, which were among the first to embrace these reforms. Over time, this new technology had a significant impact on industries such as iron, agriculture, and mining, as well as on society as a whole, resulting in increased production and changes in average living standards.
2. **2nd Industrial Revolution:** The Second Industrial Revolution, which took place between 1871 and 1914, was a period of significant development in the industrial sector. It is often referred to as the Second Technological Revolution. This era saw advancements in the telegraph and railroad industries, which allowed for the growth of infrastructure and the emergence of new ideas in the field of electricity.
3. **3rd Industrial Revolution:** This is known as the revolution of digitalization in the late 20th century, which saw a slowdown in

industrialization and technical advancements compared to previous periods. The use of human power in industrial machinery exerted significant pressure on organizations to reduce their requirements. This led to the emergence of Supercomputers, which utilize large-scale computer and communication technologies, revolutionizing the production process and driving important developments in new product technology.

The advent of the first computer marked the Industry 3.0 revolution, which was highly disruptive. For the first time, humans faced new tasks such as working with pen and paper, maintaining registers, and then uploading all the details onto computers. Currently, Industry 4.0 is characterized by new automated choices in industrial areas, eliminating the need for human intervention.

4. **4th Industrial Revolution:** It is also known as Industry 4.0 or I 4.0, which was introduced in 2011. Industry 4.0 is a contemporary framework and set of principles that revolutionize the industrial sector. The digital transformation of manufacturing and production, along with its associated processes and industries, is collectively referred to as Industry 4.0. It encompasses data models, data mapping, and value systems that span the entire product lifecycle, from introduction to workers or customers, to the business area or market, and ultimately to the value chain.

Major eight types of technology Industry 4.0:

- Vertical system integration of information system and horizontal;
- Industrial internet of things;
- The cloud;
- Augmented reality;
- Analysis of large data;
- Cyber security;
- Process of additive manufacturing; and
- Autonomous robot.

1. **Vertical System Integration of Information System and Horizontal:** In the era of Industry 4.0, there are companies that offer significant advantages and profits in networking areas. The universal integration and cross-company evolution have enabled the creation of large networks that provide fully automated, chainable value in the industry sector.

2. **Industrial Internet of Things (IIoT):** The impact of social networking is estimated to be significant, with over 100 billion linked electronic devices and a global economic value of more than $11 trillion, according to some estimates. Industry 4.0 encompasses automobiles, trucks, industrial and consumer items, as well as durable goods. These utilize components and sensors, enabling objective and powerful data analysis integrated with Internet connectivity capabilities.
3. **The Cloud:** Machine data and analytics will increase the development of the cloud, enabling the production of data-driven services for systems. As companies undertake new projects, they generate more data that can be shared across locations. Industry 4.0 is becoming more powerful, faster, and cloud computing is being adopted in industrial areas.
4. **Augmented Reality:** This message is about several related tasks that can be assisted by augmented reality. These tasks include picking parts in a warehouse, transferring items, and maintaining instructions for electronic devices. The system will provide real-time information to help make better decisions and operate more effectively in real situations.
5. **Analysis of Large Data:** It is essential to find out the hidden correlations and make informed decisions and conclusions. Nowadays, as of the current date, we analyze our data to provide answers. However, in the latest industry trends, more and more businesses are adopting slower and less traditional intelligence solutions that are efficient in big data analytics.
6. **Cyber Security:** The demand of Industry 4.0 for manufacturing industrial systems is to protect cyber security. It requires secure and sophisticated machines that rely on communication, use identity verification, access management, and other connected cyber security measures in the adoption of industrial area and communication protocols in Industry 4.0.
7. **Process of Additive Manufacture:** This term refers to the process of creating more materials by using traditional methods and manufacturing processes such as boring, milling, drilling, etc. It involves adding material as needed in an industrial setting. 3D printing is a prime example of additive manufacturing in the industry.

8. **Autonomous Robot:** Robots can communicate with humans and work alongside them in a safe environment. As time goes on, robots have a wider range of capabilities and become less expensive. An autonomous robot is capable of performing activities and exhibiting behavior without external influence from humans.

15.2.1 A PSYCHOLOGICAL IMPACT ON PEOPLE AND BUSINESS

In many industries, various sectors are generating new demands, which has significantly disrupted the established industry value on the supply side. The introduction of new technologies by innovative competitors can redefine the pricing, quality, and speed at which value is delivered, making them powerful players in the global digital platforms for development, marketing, and sales.

Businesses have recognized the need to adapt to new trends such as increased transparency, customer participation, and changing patterns of consumer activity, all of which are heavily influenced by mobile networks and data. These factors require organizations to adjust their advertising strategies, adopt the latest designs, and provide new products and services to meet the evolving demands of the market.

Four major implications business of industrial revolution:

i. Organizational forms;
ii. Enhancement of production;
iii. Innovation of collaborative; and
iv. Customer expectations.

In the market industrial area, customers are increasing in the center of the society and economic area. Whether they are enterprises or consumers, it is important to improve how we serve customers. Finally, the globalization of the industrial platform and new business models necessitate a rethinking of human nature, culture, organizational forms, and society.

The motivation of individuals is crucial amidst their busy schedules, emotions, and sometimes working independently. It is driven by their needs, desires, aspirations, and impulses in life. Nowadays, advanced technologies such as the Internet of Things, Emotional Intelligence, data science, and blockchain have become integral to the ongoing discourse on digitalization and its ability to add value.

But nowadays, human thinking and concepts are changing regarding Industry 4.0. There is a new, more recent idea emerging across the world to take advantage of the advancements in Industry 4.0 and new industrial areas. This evolution brings about changes in the paradigm of how humans and technical machines work together, known as Industry 5.0. The use of Industry 4.0 enables large-scale manufacturing improvements in production and operational excellence. This means that networks will not only keep your factories running, but also make your business more profitable in terms of discrete manufacturing and product development.

15.3 OBJECTIVES

- Find out the impact of socio-economic societies problems;
- To find out Healthcare Industry 4.0 is upcoming in 5.0 in life span.

15.4 REVIEW LITERATURE

- Alcacer & Machado (2019) found that Industry 4.0 leads to the digitalization model of business, the industrial environment, the product of industry, and industry services. All of these interconnect within a digital technology or science with virtual presentation.
- The 4th Industrial Revolution refers to the historical or previous industrial revolutions, which indicate the potential for global income level improvement and the quality of life from one person to another. It has had an impact on various aspects such as government, industry, society, and changes in daily human life (Dogaru, 2020).

15.5 GAPE

The gaps need for the research in the Ranchi District on Healthcare Industry 4.0 is upcoming in 5.0.

- There is no research on Healthcare Industry 4.0 is upcoming in 5.0.
- Studies on Healthcare Industry 4.0 are few and far between in the Indian sub-continent.
- No research on Healthcare Industry 4.0 or upcoming 5.0 has been reported on the sample from Ranchi District.

15.6 RESEARCH METHODOLOGY

- Data of personal questionnaire: This questionnaire will be prepared by the researcher. The Personal Questionnaire will be used to gather information about the subject's biodata, such as name, age, gender, religion, caste, education, and socio-economic status of the parents, etc.
- Dr. K. G. Agrawal developed the Work Motivation Questionnaire (WMQ) in 1998. This questionnaire will be used for this study.
- Socio-economic scale.
- Healthcare status scale.

15.6.1 ANALYSIS OF STATISTICAL

In this study, the data will be analyzed using a software package for statistical analysis (SPSS). A chi-square test will be conducted for categorical variables, and a t-test will be performed to assess the differences between groups.

15.6.2 FINDING

- In this study there are so many industrial healthcare problems in Indian people.
- I have found out that the people in the new industrial area are facing numerous problems in their lives.
- In this study, we find that Healthcare Industry 4.0 is upcoming, but it has some problems. We will try to solve these problems.

15.7 RESULT AND DISCUSSION

The Industry 4.0 of the 4th Industrial Revolution defines what individuals need to do. Human motivation, emotions, life security, purchasing patterns, leadership, and the allocation of time for leisure, work, and thinking all contribute to the development of our services and the cultivation of human abilities. These factors also impact the relationships between employees and workers, fostering various connections within the industrial sector.

Nowadays, this is very helpful and powerful in affecting our mental and physical health, contributing to a "quantified" self-awareness, and promoting human growth sooner than we may realize, expanding the limits of our imagination. However, with the advent of Industry 5.0, the integration of technology into people's lifestyles and the industrial workforce becomes even more powerful and inevitable. It greatly assists society by enhancing the abilities and hard work of workers in the industry, fostering good relationships between organizations and employees. A good example of this is electronic items, such as smartphones.

Industry 4.0 to 5.0 is one of the most significant challenges offered by modern information technology. People comprehend both mentally and physically why it holds such importance in our lives. Industry 5.0 introduces new networking, challenges, vast amounts of information, critical components, and the sharing of diverse industry-related information through new network connections. In the future, we will come to realize that network debates have fundamental concerns that deeply impact our inner lives.

The transition from Industry 4.0 to upcoming 5.0 indicates that the current limitations of human lifespan, both in terms of mental and physical health, are compelling us to reconsider our moral and ethical boundaries, as well as the intellectual and physical capacities of humans.

SL. No.	Variables	Percentage (%)
1.	Reduce the overall maintenance costs.	5–10%
2.	Up timing and Increase manufacturing equipment availability.	10–20%
3.	Subdue maintenance planning time.	20–50%

In these results, studies show that the analysis data are significant and have a working motivation and relationship with Industry 4.0 to 5.0. This produces positive results for owners and workers, which is very helpful for employees to deal with new products. This leads to long-lasting successful work in Industry 4.0 to 5.0.

15.8 CONCLUSION

Industry 4.0 to 5.0 refers to the collection and analysis of data and information as a means of realizing the potential of a good and smart industry, which is highly beneficial for human life and societal development. The industrial innovation of Industry 4.0 to 5.0 is characterized by rapid growth and a new collaborative environment in a connected world. It involves the analysis of large data sets, human processes, services in industrial areas, and the integration of IoT technology. This leads to a new and evolving development process. Industry 4.0 to 5.0, along with the concept of smart factories, is becoming a reality.

In future industrial areas of Industry 4.0 and upcoming 5.0, the latest organizational networking process or system is being developed. This includes machine-to-machine communication, machine learning, and the Internet of Things network. The use of smart machines would make industries more efficient and reduce waste materials. Data and optimization are crucial across the value chain, enabling machines in Industry 4.0 to 5.0 to communicate with each other and perform real-time data analysis automatically.

In the large industry, the transition from Industry 4.0 to 5.0 represents a framework and broad vision that encompasses well-defined and innovative industrial designs. This transition involves the integration of physical industrial areas with emerging digital technology, which is commonly referred to as cyber-physical systems (CPS).

Human beings strive for a secure life and have the concept of personal growth. They work in industrial areas and aim to excel in business ownership. They adapt their work patterns and continuously develop their skills and services. They also focus on improving their interpersonal relationships and understanding human behavior. The natural relationships with society are influenced by the advancements from Industry 4.0 to the upcoming 5.0.

The relevance of this study lies in its contribution to the existing knowledge and its applicability to present-day problems. The study can be conducted to explore the relationship between Industry 4.0 to 5.0 and various other variables, as well as psychological concepts such as social intelligence and motivational level.

In the present study, only the industry has been examined. The findings of the proposed study can be utilized to enhance the transition from Industry 4.0 to Industry 5.0. The outcomes of this study may have implications for future research in industrial domains.

15.8.1 IMPLICATIONS

The findings of the proposed study can be used to improve the upcoming Healthcare Industry 4.0 and its transition to Industry 5.0. The results of this study may have implications for future research in the healthcare industry, as well as for society and problem-solving.

15.8.2 LIMITATION

A limitation of this research review is the variety of types of healthcare problems in society due to COVID-19. There was no sub-typing of the healthcare Industry 4.0 in this review. Furthermore, the measurement of healthcare Industry was limited to data analysis scale, without considering socio-economic status, human lifestyle, gender, ethnicity, and mental and physical attractiveness. Measuring Healthcare Industry 4.0 is an upcoming aspect in 5.0. Finally, this literature review may not be applicable across different cultures and societies.

15.8.3 FUTURE OF THE RESEARCH

In the present study, only the healthcare industry has been undertaking Industry 4.0, with Industry 5.0 being upcoming. Therefore, a similar kind of study can be conducted in a larger sample size in the future. In the future, network IT and OT will be fully connected and secured to the highest standards, providing cheaper competitors. Industry 4.0 serves as the backbone of industrial improvement, while Industry 5.0 represents the future in our lives.

KEYWORDS

- **COVID-19**
- **healthcare**
- **Industry 4.0**
- **Industry 5.0**
- **psychological study**
- **secure life**

REFERENCES

Abedini, M., & Zhang, C., (2021). Dynamic performance of concrete columns retrofitted with FRP using segment pressure technique. *Compos. Struct.*, *260*, 113473. doi: 10.1016/j.compstruct.2020.113473. [CrossRef] [Google Scholar].

Alam, M. M., Malik, H., Khan, M. I., Pardy, T., Kuusik, A., & Le Moullec, Y., (2018). A survey on the roles of communication technologies in IoT-based personalized healthcare applications. *IEEE Access, 6*, 36611–36631. doi: 10.1109/ACCESS.2018.2853148.

Alcacer, V., & Machado, V. C., (2019). Scanning the Industry 4.0: A literature review on technologies for manufacturing systems. *Engineering Science and Technology an International Journal, 22*(3), 899–919. https://doi.org/10.1016/j.jestch.2019.01.006.

Ammarullah, M. I., Afif, I. Y., Maula, M. I., Winarni, T. I., Tauviqirrahman, M., Akbar, I., Basri, H., et al., (2021). Tresca stress simulation of metal-on-metal total hip arthroplasty during normal walking activity. *Materials, 14*, 7554. doi: 10.3390/ma14247554. [PMC free article] [PubMed] [CrossRef] [Google Scholar].

ASTM International. *Standard Terminology for Additive Manufacturing—General Principles—Terminology* (Vol. 1). International Organization for Standardization.

Austin, D., May, J., Andrade, J., & Jones, R., (2021). Delivering digital health: The barriers and facilitators to university-industry collaboration. *Health Policy Technol.*, *10*, 104–110. doi: 10.1016/j.hlpt.2020.10.003. [CrossRef] [Google Scholar].

Baker, S. B., Xiang, W., & Atkinson, I., (2017). Internet of things for smart healthcare: Technologies, challenges, and opportunities. *IEEE Access, 5*, 26521–26544. doi: 10.1109/ACCESS.2017.2775180.

Benis, A., Tamburis, O., Chronaki, C., & Moen, A., (2021). One digital health: A unified framework for future health ecosystems. *J. Med. Internet Res.*, *23*, e22189. doi: 10.2196/22189. [PMC free article] [PubMed] [CrossRef] [Google Scholar].

Bhavsar, K. A., Singla, J., Al-Otaibi, Y. D., Song, O. Y., Bin, Z. Y., & Bashir, A. K., (2021). Medical diagnosis using machine learning: A statistical review. *Comput. Mater. Contin.*, *67*, 107–125. doi: 10.32604/cmc.2021.014604. [CrossRef] [Google Scholar].

Boyes, H., Hallaq, B., Cunningham, J., & Watson, T., (2018). The industrial internet of things (IIoT): An analysis framework. *Comput. Ind.*, *101*, 6–18. doi: 10.1016/j.compind.2018.04.015. [CrossRef] [Google Scholar].

Cai, C., Wu, X., Liu, W., Zhu, W., Chen, H., Qiu, J. C. D., Sun, C. N., et al., (2020). Selective laser melting of near-α titanium alloy Ti-6Al-2Zr-1Mo-1V: Parameter optimization, heat treatment and mechanical performance. *J. Mater. Sci. Technol.*, *57*, 51–64. doi: 10.1016/j.jmst.2020.05.004. [CrossRef] [Google Scholar].

Cao, B., Wang, X., Zhang, W., Song, H., & Lv, Z., (2020). A many-objective optimization model of industrial internet of things based on private blockchain. *IEEE Netw.*, *34*, 78–83. doi: 10.1109/MNET.011.1900536. [CrossRef] [Google Scholar].

Carvalho, N., Chaim, O., Cazarini, E., & Gerolamo, M., (2018). Manufacturing in the fourth industrial revolution: A positive prospect in sustainable manufacturing. *Procedia Manuf.*, *21*, 671–678. doi: 10.1016/j.promfg.2018.02.170. [CrossRef] [Google Scholar].

Chen, Y., He, L., Li, J., & Zhang, S., (2018). Multi-criteria design of shale-gas-water supply chains and production systems towards optimal life cycle economics and greenhouse gas emissions under uncertainty. *Comput. Chem. Eng.*, *109*, 216–235. doi: 10.1016/j.compchemeng.2017.11.014. [CrossRef] [Google Scholar].

Clark, W. W., & Cooke, G., (2014). Chapter 2—The green industrial revolution. In: Clark, W. W., (ed.), *Global Sustainable Communities Handbook: Green Design Technologies and Economics* (pp. 13–39). Butter-Worth-Heinemann; Boston, MA, USA.

Cynerio, (2022). *Research Report: The State of Healthcare IoT Device Security*. Available online: https://www.cynerio.com/landing-pages/the-state-of-healthcare-iot-device-security-2022 (accessed on 11 January 2023).

Deberdieva, E. M., Vechkasova, M. V., Golikava, H. S., Borisova, A. A., & Lysenko, A. F., (2019). Experience in the management of business processes with the use of digital technologies by Russian companies of a petrochemical complex. *IOP Conf. Ser. Mater. Sci. Eng.*, *483*, 012066. doi: 10.1088/1757-899X/483/1/012066.

Dhanvijay, M. M., & Patil, S. C., (2019). Internet of things: A survey of enabling technologies in healthcare and its applications. *Comput. Netw.*, *153*, 113–131. doi: 10.1016/j.comnet.2019.03.006. [CrossRef] [Google Scholar].

Dogaru, L., (2020). The main goals of the fourth industrial revolution. Renewable energy perspectives. *Procedia Manuf.*, *46*, 397–401. doi: 10.1016/j.promfg.2020.03.058.

El-Rashidy, N., El-Sappagh, S., Islam, S., El-Bakry, H., & Abdelrazek, S., (2020). End-to-end deep learning framework for coronavirus (COVID-19) detection and monitoring. *Electronics, 9*, 1439. doi: 10.3390/electronics9091439. [CrossRef] [Google Scholar].

Emerson, N., Carre, M., Reilly, G. C., & Offiah, A. C., (2011). Geometrically accurate 3D FE models from medical scans created to analyze the causes of sports injuries. *Procedia Eng.*, *13*, 422–427. doi: 10.1016/j.proeng.2011.05.108. [CrossRef] [Google Scholar].

Faden, R. R., Kass, N. E., Goodman, S. N., Pronovost, P., Tunis, S., & Beauchamp, T. L., (2013). An ethics framework for a learning health care system: A departure from traditional research ethics and clinical ethics. *Hast. Cent. Rep.*, *43*, S16–S27. doi: 10.1002/hast.134. [PubMed] [CrossRef] [Google Scholar].

FDA, (2021). *FDA CDRH and Medical Device Cybersecurity: Response to NIST Regarding President's Executive Order (EO) on Improving the Cybersecurity of the Federal Government (EO 14028).* FDA; Washington, DC, USA. [Google Scholar].

Galal, H. *The Impact of 5G: Creating New Value Across Industries and Society*. Available online: https://www.pwc.com/gx/en/about/contribution-to-debate/world-economic-forum/the-impact-of-5g.html (accessed on 11 January 2023).

Gao, N., Luo, D., Cheng, B., & Hou, H., (2020). Teaching-learning-based optimization of a composite metastructure in the 0–10 kHz broadband sound absorption range. *J. Acoust. Soc. Am.*, *148*, EL125–EL129. doi: 10.1121/10.0001678. [PubMed] [CrossRef] [Google Scholar].

Global Market Insights Global Market Insights—Telemedicine Market. https://www.gminsights.com/industry-analysis/telemedicine-market (accessed on 11 January 2023).

Goel, S., Hawi, S., Goel, G., Thakur, V. K., Agrawal, A., Hoskins, C., Pearce, O., Hussain, T., Upadhyaya, H. M., Cross, G., et al., (2020). Resilient and agile engineering solutions to address societal challenges such as coronavirus pandemic. *Mater. Today Chem.*, *17*, 100300. doi: 10.1016/j.mtchem.2020.100300. [PMC free article] [PubMed] [CrossRef] [Google Scholar].

Hau, Y. S., & Chang, M. C., (2021). Healthcare information technology convergence to effectively cope with the COVID-19 crisis. *Health Policy Technol.*, *10*, 27, 28. doi: 10.1016/j.hlpt.2020.10.010. [PMC free article] [PubMed] [CrossRef] [Google Scholar].

He, S., Guo, F., Zou, Q., & Ding, H., (2020). MRMD2.0: A python tool for machine learning with feature ranking and reduction. *Curr. Bioinform.*, *15*, 1213–1221. doi: 10.2174/1574893615999200503030350. [CrossRef] [Google Scholar].

Hu, J. W., Zheng, B. Y., Wang, C., Zhao, C. H., Hou, X. L., Pan, Q., & Xu, Z., (2020). A survey on multi-sensor fusion based obstacle detection for intelligent ground vehicles in off-road environments. *Front. Inf. Technol. Electron. Eng.*, *21*, 675–692. doi: 10.1631/FITEE.1900518. [CrossRef] [Google Scholar].

Islam, S. M. R., Kwak, D., Kabir, M. H., Hossain, M., & Kwak, K. S., (2015). The Internet of things for health care: A comprehensive survey. *IEEE Access, 3*, 678–708. doi: 10.1109/ACCESS.2015.2437951.

Jiang, Q., Wang, G., Zhang, T., & Wang, Y., (2010). Predicting human microRNA-disease associations based on support vector machine. *Proceedings of the 2010 IEEE International Conference on Bioinformatics and Biomedicine (BIBM)* (pp. 467–472). Hong Kong, China. [Google Scholar].

Jung, Y., & Agulto, R., (2021). A public platform for virtual IoT-based monitoring and tracking of COVID-19. *Electronics, 10*, 12. doi: 10.3390/electronics10010012. [CrossRef] [Google Scholar].

Kamble, S. S., Gunasekaran, A., & Gawankar, S. A., (2018). Sustainable Industry 4.0 framework: A systematic literature review identifying the current trends and future perspectives. *Process Saf. Environ. Prot.*, *117*, 408–425. doi: 10.1016/j.psep.2018.05.009. [CrossRef] [Google Scholar].

Karami, B., Koushki, R., Arabgol, F., Rahmani, M., & Vahabie, A. H., (2021). Effectiveness of virtual/augmented reality–based therapeutic interventions on individuals with autism spectrum disorder: A comprehensive meta-analysis. *Front. Psychiatry, 12*, 887. doi: 10.3389/fpsyt.2021.665326. [PMC free article] [PubMed] [CrossRef] [Google Scholar].

Kaniyar, N. K., Goel, G., & Hawi, S., (2021). Nature-inspired materials: Emerging trends and prospects. *NPG Asia Mater.*, *13*, 5. doi: 10.1038/s41427-021-00322-y. [CrossRef] [Google Scholar].

Kiran, D. R., (2019). *Production Planning and Control. A Comprehensive Approach.* Butterworth-Heinemann; Amsterdam, The Netherlands. [Google Scholar].

Kleiven, S., & Hardy, W. N., (2002). Correlation of an FE model of the human head with local brain motion-consequences for injury prediction. *Stapp Car Crash J.*, *46*, 123–144. doi: 10.4271/2002-22-0007. [PubMed] [CrossRef] [Google Scholar].

Koptyug, A., Bäckström, M., Vega, C. A. B., Popov, V., & Chudinova, E., (2018). Developing new materials for electron beam melting: Experiences and challenges. *Mater. Sci. Forum.*, *941*, 2190–2195. doi: 10.4028/www.scientific.net/MSF.941.2190. [CrossRef] [Google Scholar].

Kurdi, H., Alsalamah, S., Alatawi, A., Alfaraj, S., Altoaimy, L., & Ahmed, S. H., (2019). Healthy broker: A trustworthy blockchain-based multi-cloud broker for patient-centered e-Health services. *Electronics, 8*, 602. doi: 10.3390/electronics8060602. [CrossRef] [Google Scholar].

Larrañaga-Altuna, M., Zabala, A., Llavori, I., Pearce, O., Nguyen, D. T., Caro, J., Mescheder, H., Endrino, J. L., Goel, G., Ayre, W. N., et al., (2021). Bactericidal surfaces: An emerging 21st-century ultra-precision manufacturing and materials puzzle. *Appl. Phys. Rev.*, *8*, 021303. doi: 10.1063/5.0028844. [CrossRef] [Google Scholar].

Lechman, E., & Marszk, A., (2019). Chapter 2—The fifth technological revolution: Context and background. In: Lechman, E., & Marszk, A., (eds.), *ICT-Driven Economic and Financial Development: Analyses of European Countries* (pp. 9–51). Academic Press; Cambridge, MA, USA. [CrossRef] [Google Scholar].

Li, B. H., Liu, Y., Zhang, A. M., Wang, W. H., & Wan, S., (2020). A survey on blocking technology of entity resolution. *J. Comput. Sci. Technol., 35*, 769–793. Doi: 10.1007/s11390-020-0350-4. [CrossRef] [Google Scholar].

Li, R., Lai, D., & Lee, W., (2017). A survey on biofeedback and actuation in wireless body area networks (WBANs) *IEEE Rev. Biomed. Eng., 10*, 162–173. Doi: 10.1109/RBME.2017.2738009.

Li, T., Xu, M., Zhu, C., Yang, R., Wang, Z., & Guan, Z., (2019). A deep learning approach for multi-frame in-loop filter of HEVC. *IEEE Trans. Image Process., 28*, 5663–5678. Doi: 10.1109/TIP.2019.2921877. [PubMed] [CrossRef] [Google Scholar].

Likens, S. *The Essential Eight Technologies That Matter Most for Business Today*. https://www.pwc.com/gx/en/issues/technology/essential-eight-technologies.html (accessed on 11 January 2023).

Liu, S., Chan, F. T., & Ran, W., (2016). Decision making for the selection of cloud vendor: An improved approach under group decision-making with integrated weights and objective / subjective attributes. *Expert. Syst. Appl., 55*, 37–47. Doi: 10.1016/j.eswa.2016.01.059. [CrossRef] [Google Scholar].

Lv, X., Li, N., Xu, X., & Yang, Y., (2020). Understanding the emergence and development of online travel agencies: A dynamic evaluation and simulation approach. *Internet Res., 30*, 1783–1810. Doi: 10.1108/INTR-11-2019-0464. [CrossRef] [Google Scholar].

Lv, Z., & Kumar, N., (2020). Software defined solutions for sensors in 6G/IoE. *Comput. Commun., 153*, 42–47. Doi: 10.1016/j.comcom.2020.01.060.

Lv, Z., & Qiao, L., (2020). Analysis of healthcare big data. *Futur. Gener. Comput. Syst., 109*, 103–110. Doi: 10.1016/j.future.2020.03.039. [CrossRef] [Google Scholar].

Lv, Z., & Qiao, L., (2020). Deep belief network and linear perceptron based cognitive computing for collaborative robots. *Appl. Soft Comput., 92*, 106300. Doi: 10.1016/j.asoc.2020.106300. [CrossRef] [Google Scholar].

Lv, Z., & Song, H., (2020). Mobile internet of things under data physical fusion technology. *IEEE Internet Things J., 7*, 4616–4624. Doi: 10.1109/JIOT.2019.2954588. [CrossRef] [Google Scholar].

Lv, Z., & Xiu, W., (2020). Interaction of edge-cloud computing based on SDN and NFV for next generation IoT. *IEEE Internet Things J., 7*, 5706–5712. Doi: 10.1109/JIOT.2019.2942719.

Marino, E., Barbieri, L., Colacino, B., Fleri, A. K., & Bruno, F., (2021). An augmented reality inspection tool to support workers in Industry 4.0 environments. *Comput. Ind., 127*, 103412. Doi: 10.1016/j.compind.2021.103412. [CrossRef] [Google Scholar].

Market Research Engine IoT in Healthcare Market Research Report. Available online: https://www.marketresearchengine.com/iot-in-healthcare-market (accessed on 11 January 2023).

Markets and Markets Research Private Ltd. *Wearable Healthcare Devices Market by Type (Diagnostic (ECG, Heart, Pulse, BP, Sleep), Therapeutic (Pain, Insulin)), Application (Fitness, RPM), Product (Smartwatch, Patch), Grade (Consumer, Clinical), Channel (Pharmacy, Online)-Global Forecast to 2025.* Markets and Markets Research Private Ltd.

Marques, G., Pitarma, R., Garcia, N. M., & Pombo, N., (2019). Internet of things architectures, technologies, applications, challenges, and future directions for enhanced living environments and healthcare systems: A review. *Electronics, 8*, 1081. Doi: 10.3390/electronics8101081.

Matrosova, E., & Tikhomirova, A., (2018). Intelligent data processing received from radio frequency identification system. *Procedia Comput. Sci., 145*, 332–336. Doi: 10.1016/j.procs.2018.11.080.

Mincholé, A., Camps, J., Lyon, A., & Rodríguez, B., (2019). Machine learning in the electrocardiogram. *J. Electrocardiol., 57*, S61–S64. Doi: 10.1016/j.jelectrocard.2019.08.008. [PubMed] [CrossRef] [Google Scholar].

Mohan, G., & Subashini, M. M., (2019). Chapter 4—Medical imaging with intelligent systems: A review. In: Sangaiah, A. K., (ed.), *Deep Learning and Parallel Computing Environment for Bioengineering Systems* (pp. 53–73). Academic Press; Cambridge, MA, USA. [Google Scholar].

Noor, N., Shapira, A., Edri, R., Gal, I., Wertheim, L., & Dvir, T., (2019). 3D printing of personalized thick and perfusable cardiac patches and hearts. *Adv. Sci., 6*, 1900344. Doi: 10.1002/advs.201900344. [PMC free article] [PubMed] [CrossRef] [Google Scholar].

North, M. M., North, S. M., & Coble, J. R., (2002). *Handbook of Virtual Environments: Design, Implementation, and Applications* (pp. 1065–1078). Lawrence Erlbaum Associates Publishers; Mahwah, NJ, USA: Virtual Reality Therapy: An Effective Treatment for Psychological Disorders. Human Factors and Ergonomics. [Google Scholar].

Periyasamy, M., Karthikeyan, S., & Mahendran, G., (2020). Electromagnetic immunity testing of radio frequency identification devices (RFID) in healthcare environment—A selected review. *Mater. Today Proc., 11*, 372. Doi: 10.1016/j.matpr.2020.09.372. [CrossRef] [Google Scholar].

Petrone, N., Candiotto, G., Marzella, E., Uriati, F., Carraro, G., Bäckström, M., & Koptyug, A., (2019). Feasibility of using a novel instrumented human head surrogate to measure helmet, head and brain kinematics and intracranial pressure during multidirectional impact tests. *J. Sci. Med. Sport., 22*, S78–S84. Doi: 10.1016/j.jsams.2019.05.015.

Petrone, N., Carraro, G., Castello, S. D., Broggio, L., Koptyug, A., & Bäckström, M., (2018). A novel instrumented human head surrogate for the impact evaluation of helmets. *Proceedings, 2*, 269. Doi: 10.3390/proceedings2060269. [CrossRef] [Google Scholar].

Ponciano, V., Pires, I. M., Ribeiro, F. R., Garcia, N. M., Villasana, M. V., Lameski, P., & Zdravevski, E., (2020). Machine learning techniques with ECG and EEG data: An exploratory study. *Computers, 9*, 55. Doi: 10.3390/computers9030055. [CrossRef] [Google Scholar].

Popov, V. V., Katz-Demyanetz, A., Koptyug, A., & Bamberger, M., (2019). Selective electron beam melting of $Al_{0.5}CrMoNbTa_{0.5}$ high entropy alloys using elemental powder blend. *Heliyon, 5*, e01188. doi: 10.1016/j.heliyon.2019.e01188.

Popov, V. V., Muller-Kamskii, G., Katz-Demyanetz, A., Kovalevsky, A., Usov, S., Trofimcow, D., Dzhenzhera, G., & Koptyug, A., (2019). Additive manufacturing to veterinary practice: Recovery of bony defects after the osteosarcoma resection in canines. *Biomed. Eng. Lett., 9*, 97–108. Doi: 10.1007/s13534-018-00092-7. [PMC free article] [PubMed] [CrossRef] [Google Scholar].

Popov, V. V., Muller-Kamskii, G., Kovalevsky, A., Dzhenzhera, G., Strokin, E., Kolomiets, A., & Ramon, J., (2018). Design and 3D-printing of titanium bone implants: Brief review of approach and clinical cases. *Biomed. Eng. Lett.*, *8*, 337–344. Doi: 10.1007/s13534-018-0080-5. [PMC free article] [PubMed] [CrossRef] [Google Scholar].

Popov, V., Lobanov, M., Stepanov, S., Qi, Y., Muller-Kamskii, G., Popova, E., Katz-Demyanetz, A., & Popov, A., (2021). Texturing and phase evolution in Ti-6Al-4V: Effect of electron beam melting process, powder re-using, and HIP treatment. *Materials, 14*, 4473. Doi: 10.3390/ma14164473. [PMC free article] [PubMed] [CrossRef] [Google Scholar].

Portnoy, J., Waller, M., & Elliott, T., (2020). Telemedicine in the era of COVID-19. *J. Allergy Clin. Immunol. Pract.*, *8*, 1489–1491. Doi: 10.1016/j.jaip.2020.03.008. [PMC free article] [PubMed] [CrossRef] [Google Scholar].

Qadri, Y. A., Nauman, A., Bin, Z. Y., Vasilakos, A. V., & Kim, S. W., (2020). The future of healthcare internet of things: A survey of emerging technologies. *IEEE Commun. Surv. Tutor.*, *22*, 1121–1167. Doi: 10.1109/COMST.2020.2973314. [CrossRef] [Google Scholar].

Qi, J., Yang, P., Min, G., Amft, O., Dong, F., & Xu, L., (2017). Advanced internet of things for personalized healthcare systems: A survey. *Pervasive Mob. Comput.*, *41*, 132–149. Doi: 10.1016/j.pmcj.2017.06.018. [CrossRef] [Google Scholar].

Qian, J., Feng, S., Tao, T., Hu, Y., Li, Y., Chen, Q., & Zuo, C., (2020). Deep-learning-enabled geometric constraints and phase unwrapping for single-shot absolute 3D shape measurement. *APL Photonics, 5*, 046105. Doi: 10.1063/5.0003217. [CrossRef] [Google Scholar].

Qiu, T., Shi, X., Wang, J., Li, Y., Qu, S., Cheng, Q., Cui, T., & Sui, S., (2019). Deep learning: A rapid and efficient route to automatic metasurface design. *Adv. Sci.*, *6*, 1900128. Doi: 10.1002/advs.201900128. [PMC free article] [PubMed] [CrossRef] [Google Scholar].

Reddy, M., (2021). *Digital Transformation in Healthcare in 2021: 7 Key Trends*. Available online: https://www.digitalauthority.me/resources/state-of-digital-transformation-healthcare/ (accessed on 11 January 2023).

Sansana, J., Joswiak, M. N., Castillo, I., Wang, Z., Rendall, R., Chiang, L. H., & Reis, M. S., (2021). Recent trends on hybrid modeling for Industry 4.0. *Comput. Chem. Eng.*, *151*, 107365. Doi: 10.1016/j.compchemeng.2021.107365. [CrossRef] [Google Scholar].

Shanbhag, G., & Vlasea, M., (2021). Powder reuse cycles in electron beam powder bed fusion—Variation of powder characteristics. *Materials, 14*, 4602. Doi: 10.3390/ma14164602. [PMC free article] [PubMed] [CrossRef] [Google Scholar].

Sherry, J., (2017). *Grand View Research Healthcare Augmented & Virtual Reality Market Worth $5.1 Billion by 2025*. Available online: http://www.grandviewresearch.com.

Shishkin, A., Goel, G., Baronins, J., Ozolins, J., & Hoskins, C., (2021). Using circular economy principles to recycle materials in guiding the design of a wet scrubber-reactor for indoor air disinfection from coronavirus and other pathogens. *Environ. Technol. Innov.*, *22*, 101429. Doi: 10.1016/j.eti.2021.101429. [PMC free article] [PubMed] [CrossRef] [Google Scholar].

Theodoridis, S., (2020). Chapter 1—Introduction. In: Theodoridis, S., (ed.), *Machine Learning: A Bayesian and Optimization Perspective* (2nd edn., pp. 1–17). Academic Press; Cambridge, MA, USA. [Google Scholar].

Von, E. M., & Von, E. W., (2020). The Digitalization of Healthcare. *HealthManagement.org J.*, *20*, 6. [Google Scholar].

Xiong, L., Zhang, H., Li, Y., & Liu, Z., (2016). Improved stability and H∞ performance for neutral systems with uncertain Markovian jump. *Nonlinear Anal. Hybrid Syst.*, *19*, 13–25. Doi: 10.1016/j.nahs.2015.07.005. [CrossRef] [Google Scholar].

Xiong, Z., Xiao, N., Xu, F., Zhang, X., Xu, Q., Zhang, K., & Ye, C., (2021). An equivalent exchange-based data forwarding incentive scheme for socially aware networks. *J. Signal Process. Syst.*, *93*, 249–263. Doi: 10.1007/s11265-020-01610-6. [CrossRef] [Google Scholar].

Yang, J., Li, S., Wang, Z., Dong, H., Wang, J., & Tang, S., (2020). Using deep learning to detect defects in manufacturing: A comprehensive survey and current challenges. *Materials, 13*, 5755. Doi: 10.3390/ma13245755.

Zhang, A., GoodSmith, M., Server, S., Uddin, S., McNulty, M., Sherer, R., & Lio, J., (2022). Providing support in a pandemic: A medical student telehealth service for ambulatory patients with COVID-19. *Healthcare, 10*, 100612. Doi: 10.1016/j.hjdsi.2022.100612. [PMC free article] [PubMed] [CrossRef] [Google Scholar].

Zhao, J., (2020). Neural network-based optimal tracking control of continuous-time uncertain nonlinear system via reinforcement learning. *Neural Process. Lett.*, *51*, 2513–2530. Doi: 10.1007/s11063-020-10220-z. [CrossRef] [Google Scholar].

Zhu, Q., (2020). Research on road traffic situation awareness system based on image big data. *IEEE Intell. Syst.*, *35*:18–26. Doi: 10.1109/MIS.2019.2942836.

CHAPTER 16

Analysis of Sustainable Development in Spatial Politics in the Construction of the Raknamo East Nusa Tenggara in Indonesia

MELKISEDEK N. B. C. NEOLAKA,[1] DIANA SAN TABUN,[1] and UMA SHANKAR YADAV[2]

[1]*Faculty of Social Science and Political Science, University of Nusa Cendana, NTT, Indonesia*

[2]*Department of Humanities and Social Science, Motilal Nehru National Institute of Technology, Uttar Pradesh, India*

ABSTRACT

This research focuses on the political perspective of community empowerment in development policy. The aim is to encourage efforts that reconcile the interests and orientations of each actor in an equal and intensive manner, resulting in accommodative and participatory outcomes that satisfy all interacting actors. The analysis reveals that the orientation tendency of each actor is not absolute but allows for a shift in direction based on the actor's orientation results. This shift leads to the accommodation of political interests for each actor. The pattern of interaction aims to achieve policy results that accommodate the interests of all actors involved. As a consequence, this human-oriented development model highly values the existence of individuals, both as individuals and as members of local communities. It emphasizes local initiatives, self-organization through

Sustainable Strategic Business Infrastructure Development and Contemporary Digital Practices in Industry 5.0. Indrajit Ghosal, Bikram Prasad, Nilanjan Ray, & Lalit Kumar Joshi (Eds.)

organizational units on a human scale, and independent communities. The model recognizes the importance of economic efforts in achieving prosperity and self-realization. It acknowledges that participating in economic activities not only contributes to one's self-esteem, but also enhances the quality of life through consumption of products.

The facts on the ground include the emergence of various phenomena such as marginalization, unidimensionalization, massification, and deprivation. These phenomena arise amidst the pursuit of economic growth through the implementation of the growth paradigm. Based on these facts, proponents of alternative development theory are optimistic that the human development paradigm can prioritize values such as self-esteem, respect, authenticity, dignity, and self-actualization as the foundation for development.

16.1 INTRODUCTION

From an elite perspective, the politics of development in Indonesia, both in the Old Order and New Order eras, greatly disregarded the role of society as development actors in the process of planning, implementation, and efforts to sustain development outcomes. During the Old Order era, politicking occurred in all sectors, causing development, particularly the improvement of people's welfare, to be overshadowed by ideological and sensational political issues. Policies continue to be dominated by the orientation of political actors in the context of shaping and consolidating nationalism.

During the New Order era, the development policies were oriented towards achieving economic growth that could catch up with other countries. These policies were implemented in a centralized manner by the government, particularly the central government. The entire national potential was mobilized to achieve optimal growth.

Repressive measures, especially concerning mass participation, were taken to maintain socio-political stability, ensuring that development could be carried out optimally. The role of the community as a development actor was consistently sacrificed in favor of development and other larger interests, specifically national economic growth.

The orientation of government actors is aimed at achieving development effectiveness, with a strong assumption that the community lacks sufficient technical knowledge and skills. The preparation and implementation of development projects often exclude community participation, and when involved, it is mainly for the sake of legitimacy.

This development policy paradigm heavily emphasizes economic measures as the sole indicator of success. This approach is common in many developing countries, where the pursuit of economic growth is prioritized in development goals. In this capitalistic policy model, development is reduced to mere "economic growth" (Sachs, 1995). Consequently, development is primarily focused on increasing income levels in economically disadvantaged areas.

In this "economic development" model, the central factor to achieve maximum growth is the capital factor. Overall, this model formulation does not consider social issues such as poverty, income distribution, employment, and the impact of technology used in the production process. Additionally, the theoretical framework put forward also does not consider aspects of orientation and community institutions.

16.1.1 SIGNIFICANCE OF THE CHAPTER

This chapter is helpful for policymakers and members of society who are interested in sustainable development, politics, and management. It provides a clear understanding of the basic concepts and ideas related to human development. Many people have come to realize that the prevailing development paradigm, which prioritizes economic growth, places excessive emphasis on "production" rather than its impact on human well-being. This chapter emphasizes the importance of prioritizing "human" in relation to production, rather than the other way around.

Policymakers believe that with this trickle-down, a harmonious process will occur in society, spreading the benefits of economic growth to all strata of society through a mechanism known as the trickle-down effect. So, the issue of people's welfare is essentially a matter of time. As long as people are willing to be patient and not demand too much, they will eventually receive a gradual increase in welfare as a logical consequence of the implemented development (economic growth).

The logical consequence of this "development economic" model is to disregard the involvement of community actors in the preparation, implementation, and benefits of development. Not only are communities not included in the development process, but they are also excluded from it and do not enjoy its results. As a result, numerous social and economic problems arise as soon as development is planned and carried out, continuing even after the completion of the development program.

16.2 LITERATURE REVIEW

The "development economic" approach in the realm of development theory is also known as the structuralist approach (Arief & Sasono, 1990; Baran, 1957; Sachs, 1995). The structuralist approach is based on the reality that occurred in Latin American countries, with a theoretical exposition that rests on the process of capital accumulation and socio-economic changes that occurred in developing countries, the majority of which were former colonies. In the entire economic system of the former colonial countries, a highly unequal pattern of income distribution and distribution of control over economic resources has finally been formed. The structure of society consists of a small group of high-income residents as compradors of the elite in developed countries and the mass of developing people.

In the context of policy formulation, the interaction between the government's orientation, community actors, and other actors leads to structural inequality. This inequality arises from the pursuit of broader interests, primarily those of international capitalism, the center over the regions, cities over villages, and the rich over the poor. As a result, the formulation and implementation of policies are structurally controlled.

This structuralist paradigm has proven to be incapable of addressing the development challenges that benefit the people. As a result, a new critique has emerged in the field of development theory, known as the neo-structuralist paradigm. The neo-structuralist development paradigm emphasizes the importance of fostering a socio-cultural system that generates high-quality, resilient, and socially conscious human capital. A key aspect of development policy is the involvement of the community in the planning and implementation processes. This approach aims to empower the community and integrate their perspectives in decision-making and the execution of development programs.

Development policies in Indonesia demonstrate that by disregarding the community aspect (lack of participation) in the development process, it fails to achieve maximum output. The outcomes of development primarily consist of physical objects that are disconnected from the needs and presence of the surrounding community.

This non-participatory development process sacrifices the interests of the community more than the benefits felt by the community. Development policies like this can lead to public apathy and even antipathy in the long term, as shown by an indifferent attitude towards the preservation of development results and a tendency to damage them.

Previous experience in water resources management policies in Indonesia has demonstrated that the orientation and interests of the community are significantly marginalized in all policy processes. Therefore, it is necessary to change the orientation of water resources management policies to be more participatory. This will ensure that development, ideally oriented towards the interests of the people, is accepted and supported by the people due to its participatory nature and lack of harm inflicted upon them.

Water resources management policies in various cases and in various countries show the fact that society is often the object and victim of development. Especially in the development of water resources infrastructure, which is oriented towards economic growth. Development policies of this kind always require enormous social sacrifices, not only in the process of preparation and implementation but also in the process of preserving them.

16.2.1 POLICY IMPLEMENTATION

In the public policy process emphasizes the importance of policy implementation, which is considered even more crucial than Udoji's policy making. In Abdul Ekedar, policy implementation involves navigating conflicts, making decisions, and determining who benefits from a policy. It serves as a bridge between policy formulation and the desired outcomes. According to Anderson (in Islamy, 2000, p. 68), there are four aspects that require examination in policy implementation: the actors responsible for implementation, the nature of the implementation process, compliance, and the impact of policy implementation.

Based on the understandings of policy implementation above, it can be concluded that policy implementation is an activity in implementing policies to distribute policy results by policy implementers to target groups in order to achieve the results and objectives of the policy. The results and objectives of a policy will certainly not be achieved by itself without the policy being implemented. Although implementation is known as a concept, it has a very deep purpose and meaning to describe the efforts made by implementers in realizing policies, because the success of policy implementation is closely related to various aspects including consideration of policy makers, commitment, and consistency of implementers, policies, and target behavior. The policy implementation process is part of the policy process from the perspective of social and political change.

This model assumes that policy implementation runs linearly from the available political decisions, implementation, and performance of public policies.

The top-down approach model formulated by Van Metter and Van Horn is called the A model of Policy Implementation. This implementation process is an abstraction or performance of policy implementation, which is deliberately carried out to achieve high performance of public policy implementation in relation to various variables. Van Metter and Van Horn emphasize the importance of differentiating policy content because the effectiveness of implementation will vary greatly depending on the type and issue of the policy. The factors that influence the implementation process will also be very different. According to them, the type of policy will require specific characteristics in terms of process, structure, and relationship between various factors in its implementation.

1. **Policy Size and Purpose:** The success of policy implementation can be measured by the extent to which the size and objectives of the policy align with the existing socio-cultural context at the implementation level. When the size or objectives of a policy are too idealistic or utopian to be implemented by citizens, it becomes challenging to achieve success in realizing public policies.
2. **Resources:** The success of the policy implementation process is highly dependent on the ability to utilize the available resources. Humans are the most important resource in determining a successful implementation process. Certain stages of the entire implementation process require qualified human resources in accordance with the work required by politically determined policies. However, when the competence and capability of these resources are lacking, it becomes difficult to expect the performance of public policies.
3. **Characteristics of Implementing Agents:** The focus of attention on implementing agencies includes formal organizations and informal organizations involved in implementing public policies. Van Metter and Van Horn considered that by looking at the characteristics of

implementing agencies, it could not be separated from the bureaucratic structure. Van Metter and Van Horn describe several elements that may affect an organization in implementing policies: (i) Competence and size of an agency's staff; (ii) hierarchical level of supervision of sub-unit decisions and processes in implementing agencies; (iii) the political resources of an organization (e.g., support among members of the legislature and executive); (iv) the vitality of an organization; (v) the level of "open" communications defined as a network of independent horizontal and vertical communications and a relatively high degree of freedom in communication with individuals outside the organization; and (vi) formal and informal links within a body with decision-making or decision-making bodies.

4. **Attitudes or Tendencies (Disposition) of the Implementers:** The attitude of acceptance and rejection of the implementing agent will greatly affect the success or failure of the performance of public policy implementation.
5. **Communication between Organizations and Implementing Activities:** Implementation will run effectively if the measures and objectives are understood by the individuals who are responsible for policy performance (In Agustino, 2017, p. 135). Coordination is a mechanism as well as the main requirement in determining the success of policy implementation.
6. **Economic, Social, and Political Environment:** Van Metter and Van Horn propose to consider the following questions regarding the economic, social, and political environment that affects the jurisdiction or organization in which implementation.

16.3 METHODOLOGY

This study utilizes the narrative review method to consolidate the findings of multiple studies from reputable journals, reviews, and chapter books (Popay et al., 2016).

16.3.1 RESEARCH APPROACH AND FOCUS

Based on the subject matter studied, this research utilizes a qualitative approach. Qualitative research has main characteristics that are concerned with meaning, context, and a research process that is more cyclical than linear. Data collection and data analysis occur simultaneously, with a focus on depth rather than breadth of research. Additionally, the researcher themselves serves as the main instrument.

16.3.2 INFORMANTS AND RESEARCH INFORMANTS DETERMINATION TECHNIQUES

In determining the informants, the researcher uses purposive sampling, namely selecting data sources based on specific considerations. For example, the researcher may choose informants who are deemed to possess the most knowledge about the subject of interest, or informants who are policymakers, as this facilitates the exploration of the object under study.

16.3.3 DATA COLLECTION TECHNIQUES

To address the research problem, it is essential to gather a substantial amount of data and information regarding the topic under discussion in this study. The data collection methods employed in this study are as follows:

1. **Field Study (Primary Data):** It is the data obtained directly from research subjects by wearing measurement tools or data collection tools directly on the subject as the sought-after information (Saifuddin Azwar, 2004, p. 91).
2. **Literature Study (Secondary Data):** It is second-hand data obtained through other sources, not directly obtained by the researcher from the research subject (Saifuddin Azwar, 2004, p. 91). Literature study (document) is a form of data collection through legal documents, literature books, articles, previous research findings, and other documents that can provide the necessary information for researchers in their research (Syamsudin, 2007, p. 101).

16.3.4 RESEARCH DATA SOURCES

What is meant by research data sources are the subjects from which the data were obtained (Arikunto, 2006, p. 129). According to Lofland &

Lofland (1984, p. 47), the main data sources in qualitative research are words and actions, while additional data such as documents and others are also considered (Moleong, 2002, p. 112):

1. **Primary Data Sources:** These data sources were obtained directly from research subjects using interview techniques. In the implementation of this technique, the authors collect raw data and manage the data.
2. **Secondary Data Sources:** This data source is obtained from books and documents available at the Mota'ain Cross-border Post and the Law of the Republic of Timor-Leste, the successful implementation of the visa-free policy, and the actual conditions of the research location. This secondary data is expected to contribute to a broader discourse and provide insights for researchers, resulting in more accurate research results.

16.3.5 DATA ANALYSIS TECHNIQUES

Qualitative data analysis, according to Bogdan and Biglen, is an effort made by working with data: organizing, sorting, and synthesizing it to find patterns, determine importance, and make decisions. This information can then be shared with others (Lexi, 2006, p. 248). Secondary data has been used and inductive content analysis has been applied for result and discussion of chapter.

16.4 RESEARCH RESULTS AND DISCUSSION

16.4.1 LOCAL COMMUNITY EMPOWERMENT

The results of development are not always in favor of local communities, especially those who are unable to compete or are marginalized in the process or in enjoying the results of such development. The results of research conducted by Choy on communities around the Bakun Dam in Malaysia concluded that the construction of a dam should be a development that enhances the socio-economic and cultural sustainability of the indigenous people around it. However, what has happened is that society has been uprooted from its historical socio-economic and ecosystem roots as a result of development being carried out in an authoritarian and repressive manner. Dam construction should be a development that protects and preserves indigenous communities with all the increasing socio-cultural and economic activities, not displacing them.

In Indonesia, the construction of dams, which technically requires the sacrifice of people's land to become inundated areas, always leaves social problems in the future. The resettlement process is detrimental to the people, breaking traditional communication and transportation routes, and even severing the community's cultural relationship with the tombs and relics of their ancestors as a result of being cut off by the inundation caused by the dam.

In many cases, the impact of dam construction also alters or eliminates the traditional economic activities of the community, typically centered around agriculture and other subsistence sectors. The relocation of people to new lands (resettlement), while geographically similar to their previous place of residence, fails to recreate the original pattern of economic and social relations.

The process of constructing dams always emphasizes technical-economic factors, which are considered the most effective and efficient way to stimulate economic growth, even though it may overlook the human factor and its associated social problems. In this development process, the alignment between government orientation and community orientation throughout the policy process is deemed unimportant. Often, community orientation needs to be avoided as it can hinder the effectiveness and efficiency of the development process. The social impact remains a ticking time bomb, even though the construction of the dam took place many years ago.

A paradigm shift in the preparation, implementation, and management of dams as a significant resource needs to be carried out, considering the social impacts that have occurred. This change should involve mutually beneficial sharing between interested stakeholders. Community involvement is an effort to change the policy paradigm, emphasizing the importance of actor orientation in the development policy process.

Development policies that incorporate community participation at every stage of the water resource management process are considered more humane alternatives. This effort aims to empower the community in managing water resources, ensuring that the preparation, implementation, and outcomes of development are guided by actors (both the state and society) who have equal positions and shared responsibilities.

The importance of the community's role in the development process in the management of water resources is emphasized by Sinclair (2002), who highlights the significance of "public consultation" in the planning, implementation, and management of the Manitoba Dam construction in Brazil. The "Manitoba Approach" model concludes that community consultation (meetings to capture and understand community perspectives) is an integral part that should be conducted at every stage of development, including preparation, implementation, and preservation of development outcomes.

Empowerment signifies humanization, specifically the independent efforts of empowered individuals to achieve self-empowerment. This condition needs to be fostered to enable people to have creativity and initiative in their own development. In other words, development should be centered around people, with individuals acting as key players in setting goals, controlling resources, and directing processes that impact their lives (Korten, 1988).

The concept of empowerment is actually more oriented towards efforts from below, namely businesses carried out by the community itself, in addition to awareness from above (the government) that the community needs to be positioned as an independent actor in the process of preparation, implementation, and preservation of development results.

In the context of development policies, especially in the management of water resources, empowerment is connoted more as an effort to create self-sufficiency among the community in order to carry out upward mobility. This mobility is especially in making decisions about its orientation and future dealing with other actors such as the state, society, or other private actors in the making, implementing, and preserving development programs. With this empowerment, it is hoped that the community will have creativity and productivity as well as initiatives with self-potential capital and control of existing territories, to achieve alignment of positions in orientational interactions between actors which usually lead to efforts to increase economic and social welfare or vertical economic and social mobility.

Policies that support the empowerment process must consider the community's orientation, the suitability with the condition of its natural resources, and prospects for future development. Communities are not only seen as objects of development, but also as development actors who have the ability to actualize and become empowered actors in the interface with the Government and other development actors.

Has the policy for developing water resources infrastructure in Indonesia, such as dams, empowered the surrounding community in its formulation? What should be the involvement of the community as one of the actors with an orientation that deals with the orientation of other actors in the process of formulating water resources infrastructure development policies? This study aims to address this focal problem.

Historically, the construction of the Raknamo Dam, which is the subject of this research study, did not involve the participation of the surrounding community. Development policies were prepared unilaterally without considering the perspectives of the community. Based on the results of interviews with the affected communities, during the dam construction process, there were instances of forceful evictions and relocations of people's settlements, disregarding the preferences of the individuals due to the strong influence of the government at that time. In addition to displacing people from their land, the construction of the dam has had implications for changing the livelihoods of the local population.

The centrally planned dam construction has also disrupted the previous ecological system. The lack of community participation in the technical design, implementation, and management of these functions has made the community feel disconnected from the existence and purpose of the dam. As a result, there is a disjointed relationship between the community and the dam, which is the outcome of a large-scale development process.

The relationship between the community's involvement, their feelings, and self-actualization, in maintaining the sustainability of the Dam, is diminished or even non-existent. This condition leads to a decline in the community's concern for the quality of the dam, and unconsciously, the community has engaged in activities that are detrimental to the conservation and preservation of the dam and its surrounding land.

From a development policy perspective, the construction of dams, such as the Raknamo Dam, ideally serves as a strategic program aimed at improving the welfare of the community, especially the farmers residing around the dam who rely on it for their agricultural cultivation. However, in reality, the benefits of the dam as a source of irrigation are enjoyed by people who are relatively far from the dam, rather than those directly affected by its existence in the vicinity.

The community residing around the Raknamo Dam has traditionally been engaged in forest-fringe farming, supplemented by part-time jobs collecting wood and fruit from the forest. However, with the construction of the Raknamo Dam, accessing the outskirts of the forest beyond the Dam has become challenging due to the distance and the need to cross or walk along the Dam's edge, which further extends the journey in a circular manner.

Consequently, many community members are seeking alternative livelihoods that may not always align with the interests of the dam's existence. These activities include collecting and selling various goods obtained from the Dam's vicinity, utilizing the upstream river of the Dam to transport illicitly acquired wood, and cultivating crops in the inundated areas of the Dam that are drying up for the sake of efficiency.

These activities have the potential to damage dams, especially by increasing sedimentation and pollution. The behavior of people around the dam tends to neglect its maintenance and care as a development asset, which ultimately shortens the dam's lifespan despite the long and expensive construction process.

To illustrate, during the initial construction phase, the Raknamo Dam had sufficient water capacity to meet the needs of the surrounding community. However, after completion, the dam's water capacity is now 13.14 million m^3, with an area of 841 Ha to be irrigated and a raw water flow of 100 liters per second. The reduced productive age of the dam is primarily caused by high sediment deposits, influenced by several factors. Firstly, the slope of the land located to the south, west, and east of the Raknamo Dam contributes to significant erosion during rainfall.

Secondly, the destruction of the buffer forest at the top of the dam, especially in the watershed that supplies the dam's water, is a concern. Forest destruction resulting from illegal logging leads to a high erosion rate upstream of the dam and accelerates sedimentation in the inundation area.

Thirdly, the improper use of the dam by the surrounding community is also an issue. Activities such as planting rice and crops in the dam's inundation area, fish farming in the inundation area, and other efforts made by the community to sustain themselves by utilizing the dam are not in accordance with the dam's operational standards. This includes unauthorized logging and the destruction of dam equipment for sale.

Community behavior that is contrary to the standard operating procedure (SOP) for the management of the Raknamo dam not only causes damage to the dam's facilities and infrastructure, but also endangers the lives of the surrounding community. The function of the dam as a source of irrigation diminishes, and the careless use of dam facilities and infrastructure accelerates the silting of the dam. The most dangerous impact is the potential bursting of the embankments around the dam, which could result in the flooding of the villages nearby.

Unfortunately, the public tends to understand this situation apathetically, as they perceive the existence of the dam as having no relevance to their lives and even consider it detrimental due to its historical construction process.

16.5 CONCLUSION

This program is interesting to examine in more depth for several reasons. First, from a pessimistic perspective, the program can be interpreted as a lip service effort by the government to gain sympathy from the public. It is evident that the government is no longer able to demonstrate its might in maintaining development results without involving the community. This, in turn, dispels the myth that the government is an all-knowing and versatile superbody in the preparation, implementation, and maintenance of development results.

From an optimistic perspective, this can be interpreted as a new paradigm in development policies in Indonesia. Despite being late, this program offers new hope for humanizing the community in the development preparation process.

Secondly, the implementation of community empowerment programs so far has been more politically strategic in nature, rather than a paradigm model of development adopted by the Government. Most community empowerment programs are carried out within the framework of a "Social safety net," so the actualization is far from the reality of what community empowerment should be.

In contrast to community empowerment programs in general in Indonesia, the Savings-Based Housing Financing Assistance (BP2BT) program is not within the framework of JPS, but it is an interactive policy formulation program developed jointly between the Government and the community to manage and maintain the sustainability of the Raknamo dam, while also being able to improve community welfare.

Third, from a policy perspective, the process of actor orientation in this program describes a human-oriented (people-oriented) community development policy model. This development model is considered more humane and not just economy-oriented.

This chapter examines how a policy is made through a process of actor orientation/elite interaction in order to produce a policy that can be considered a combination of the orientations of all actors. It is implemented and becomes the responsibility of all the actors involved.

KEYWORDS

- **data analysis**
- **data collection techniques**
- **local community empowerment**
- **policy implementation**
- **research data sources**
- **spatial planning**
- **sustainable development**

REFERENCES

Adams, A., (2000). *Social Impact of An African Dam: Equity and Distributional Issues in the Senegal River Valley* (pp. 23, 24). Independent Expert, Senegal. https://www.scirp.org/(S(ny23rubfvg45z345vbrepxrl))/reference/referencespapers.aspx?referenceid=3092562 (accessed on 11 January 2023).

Adams, W., (2000). *Downstream Impacts of Dam* (pp. 3–6). University of Cambridge, UK, https://pubs.usgs.gov/pp/1286/report.pdf (accessed on 11 January 2023).

Agustino, L., (2006). *Dasar-Dasar Kebijakan Publik.* Alfa Beta, Bandung.

Almond, G., & Dan, P. B. G., (1966). *Comparative Politics: A Developmental Approach.* Boston, Little Brown, and Company. https://www.scirp.org/reference/referencespapers?referenceid=2583638.

Arief, S., & Adi, S., (1990). *Ketergantungan dan Keterbelakangan di Indonesia*, Jakarta, Lembaga Studi Pembangunan.

Baran, P. N., (1957). *The Political Economy of Growth* (pp. 6, 7). New York, Monthly Review Press.

Batten, T. R., (1960). *Communities and Their Development.* Oxford University Press, London.

Bhatacharya, J., (1972). *Administrative Organization for Development.* IIAS, Brussel.

Bhatnagar, D., Ankita, D., Magui, M. T., & Prameeta, K., (2000). *Water Supply and Sanitation for Low Income Communities: Indonesia* (pp. 1–9). The World Bank.

Blair, H. W., (1981). *The Political Economy of Participation in Local Development Programs: Short Term Impasse and Long-Term Change in South-East Asia and the United States From the 1950s to the 1970s.* Cornell University.

Bryant, C., & White, L. G., (1982). *Managing Development in the Third World.* Westview Press, I Boulder, Colorado.

Burby, R. J., (2003). Making plans that matter: Citizen involvement of the American planning. *Journal of the American Planning Association* (Vol. 69, No. 1, pp. 33–50). Winter.

Carino, J. K., (2000). *Dams Indigenous People and Vulnerable Ethnic Minorities: A Case Study on the Ibaloy People and the Agno River Basin, Province of Benguet, Philippines* (pp. 9–13). Cordillera Peoples Alliance, Philippines. https://www.populationenvironmentresearch.org/node/8010 (accessed on 11 January 2023).

Dube, S. C., (1964). Bureaucracy and nation building in transitional society, Dalam. *International Social Science Journal, XVI*(2), 6, 7.

Dunn, W., (1995). *Analisa Kebijaksanaan Publik.* Hanindita Offset, Yogyakarta.

Ferradas, C., (2000). *Report of Social Impacts of Dams: Distributional and Equity Issues-Latin American Region* (pp. 10–15). State University of New York at Binghamton, USA.

Gudeman, S., (1986) *Economic as Culture: Models and Metaphors of Livelihood.* London and New York, Routledge, and Kegan Paul.

Hammarskjold, D., (1975). *What Now? Another Development?* Uppsala, Foundation.

Irfan, I. M., (1997). *Prinsip-Prinsip Perumusan Kebijakan Negara.* Bumi Aksara, Jakarta.

Katz, S. K., (1965). *Guide to Modernizing Administration for National Development.* GSPIA, University of Pittsburg.

Mostert, E., (2000). *Dams on Transboundary Rivers* (pp. 1–3). Delf University of Technology, Netherlands.

Ndraha, T., (1985). *Peranan Administrasi Pemerintahan Desa Dalam Pembangunan Desa.* Desertasi, Universitas Gadjah Mada, Yogyakarta.

CHAPTER 17

The Impediments to Student Life Imposed by COVID-19 in India and Bangladesh: An Empirical Analysis

DHANONJOY KUMAR,[1] NIRMAL CHANDRA ROY,[2] and NILANJAN RAY[3]

[1]*Department of Management, Islamic University Kushtia, Bangladesh*

[2]*Department of Business Administration (HR), The University of Burdwan, West Bengal, India*

[3]*Associate Professor of Marketing Management Department of Management Studies, JIS University, West Bengal, India*

ABSTRACT

The present study analyzes the impacts of COVID-19 on students' education life to identify the challenges faced by educational institutions due to the COVID-19 pandemic in India and Bangladesh.

The research utilized a descriptive survey approach, employing a structured survey questionnaire and interactive interviews to collect data. A total of 804 respondents from India and Bangladesh participated in the study, with 402 from each country. The respondents were categorized as students, teachers, and guardians. Non-probability convenience sampling and structured questionnaires were used for data collection. The measurement properties were evaluated using PLS-SEM software, and the study hypotheses were tested.

Sustainable Strategic Business Infrastructure Development and Contemporary Digital Practices in Industry 5.0. Indrajit Ghosal, Bikram Prasad, Nilanjan Ray, & Lalit Kumar Joshi (Eds.)

The study found that the uncalled-for vacations during the COVID-19 pandemic situation caused severe distractions to education, hampered regular routine learning, and changed lifestyles. A significant number of students have become addicted to social media, internet gaming, and engaged in illegal activities. Some students have even left their student life behind, opting to engage in family life, work, and other social activities. However, these circumstances present great challenges for students' educational life and educational institutions.

Now, the burning question is how to address and overcome the challenges posed by the COVID-19 pandemic. The present study focuses on this issue as its central theme.

17.1 INTRODUCTION

The COVID-19 outbreak has compelled almost all countries around the world to face disruptions and instability in their regular way of life, forcing them to adapt to the new global reality. Both India and Bangladesh have encountered similar challenges, but they have shown great courage in confronting the situation and have successfully managed the impact of the COVID-19 pandemic on various aspects of people's lives, including social, economic, health, and education. Bangladesh and India are neighboring countries, with Bangladesh being surrounded on three sides by Indian states. It shares borders with West Bengal to the west and north, Assam to the north, Meghalaya to the north and northeast, and Tripura and Mizoram to the east. The people living on both sides of the border have a long history of shared language, culture, and lifestyle. Therefore, their experiences in dealing with the waves of the COVID-19 pandemic may exhibit some similar characteristics.

In India, on March 16, 2020, the government declared a nationwide lockdown. The University Grants Commission directed universities on March 19, 2020, to postpone exams until March 31, 2020 (UNICEF, 2020). The board exams conducted by ICSE and CBSE boards were initially deferred until March 31, 2020, and later until July 2020 (The Hindu, 2020). On March 8, 2020, the first COVID case was diagnosed in Bangladesh (IEDCR, 2020). WHO declared the coronavirus as a global pandemic on March 11, 2020. The Bangladesh Government took initiatives to close educational institutions and other offices afterward. However, on March 17, 2020, the authorities decided to close all educational institutions and offices. In terms of the education system, India has been the worst sufferer between the two due to COVID-19. Beyond the shocking impact on human life, COVID-19 has significantly interrupted access to education in the country. Approximately 247 million primary and secondary school students have reportedly been out of school in India (Roy, in press). Besides CBSE and ICSE exams, many other exams like Competitive, CA, etc., also got postponed. Online learning was encouraged during the pandemic, but the fact remains that only a quarter of households in India had internet access. By April 2021, most states had opened educational institutions. Thus, COVID-19 has affected the education system in the country. COVID-19 has changed the conventional education method to the educational technologies model, where teaching, learning, and assessments are conducted online (Joshi et al., 2021).

Thus, the present study aims to identify the effect of COVID-19 on students' education life, social life, and regular education.

17.2 LITERATURE REVIEW

On January 30th, 2020, the WHO declared the outbreak of the Corona Virus a Public Health Emergency of International Concern. Subsequently, on March 11th, 2020, the WHO upgraded the COVID-19 outbreak to a global pandemic (WHO, 2020). The COVID-19 crisis has exacerbated an existing catastrophe in the education system, making disparities more visible (UNESCO, 2021), and leading to systemic changes in the global education system (Ainscow, 2005; Ainscow & Giorgis, 1999).

The current global pandemic has severely impacted the education system. Due to the emergence of COVID-19, education, which is a crucial factor for a nation's individual growth and welfare, has been jeopardized. Millions of children worldwide have been affected (Rodricks, 2021). COVID-19 is also disrupting the activities of agriculture and the agro-economy of Bangladesh (Kumar et al., 2020), and has had numerous effects on the economy and the banking industry in the country, resulting in an increase in non-performing assets (NPA), investments, and more (Kumar et al., 2021).

The COVID-19 pandemic has caused the largest disruption to the education system in human history. It has affected approximately 1.6 billion students in over 200 countries. As a result of the pandemic, schools, colleges, and other educational institutions have been forced to close (Pokhrel & Chhetri, 2021). These closures have impacted about 94% of the world's learner and student population. As of January 2021, approximately 825 million pupils have been affected by school closures in response to the pandemic (UNESCO, 2021). There has been a pressing need to introduce alternative educational and assessment strategies. However, digital teaching and learning have been implemented in all affected countries around the world (Dhawan, 2020). Initially, in most countries, administrators decided to temporarily close educational institutions to control the rate of infections to some extent. Later, in many countries, educational institutions were reopened for a few grades only, which led to an increase in the number of infections. Subsequently, they were closed again (Suresh, 2021). The COVID-19 pandemic has necessitated emergency responses and changes in the educational system. These changes have focused on various modes of emergency remote schooling (Vuorikari, Velicu, Chaudron, Cachia, & Di Gioia, 2020). UNESCO has recommended the implementation of distance learning in response to global school and educational institution closures (UNESCO, 2020). As the pandemic continues, educational institutions may continue to rely on remote learning systems. Strategies such as online teaching and learning, rotating schedules, and cohorts are being implemented to minimize physical contact and prevent the spread of COVID-19 (CDC, 2020).

Educational institution closures have highlighted various concerns that disrupt access to education, as well as socio-economic challenges such as high social and economic costs (UNESCO, 2020). The impact is particularly significant for disadvantaged students and their families, including interrupted learning, compromised nutrition, childcare problems, and subsequent economic burdens for families who are unable to work (The Telegram, 2020). This situation is truly unfortunate for students, especially those from underprivileged backgrounds, who may lack parental support and need to learn independently (OECD, 2020). In many developing and underdeveloped countries, parents are more likely to lose their jobs when schools close in order to take care of their children, resulting in loss of wages (UNESCO, 2020).

As of now, online teaching and learning have become critical for the sustenance of the education sector across the globe, in order to avoid community transmission of the coronavirus (Skulmowski & Rey, 2020). Modern technology enables students and teachers to access specific materials beyond textbooks, in various formats, reducing time and space constraints (OECD, 2020). Online teaching and learning can take various forms, including automated solutions like the latest learning tools, or low-tech alternatives such as television, radio, SMS, etc. (Murphy, 2020). The majority of educational institutions worldwide have introduced online teaching and learning through video telephony tools such as Google Meet, Google Classroom, Zoom, etc. However, in rural areas, especially in developing and underdeveloped countries, and among students from disadvantaged families, limited internet access hinders their ability to learn online (OECD, 2020). There has been a negative impact on learning outcomes in online education (UNESCO, 2020). While educational institutions offer necessary learning resources, the closure of these institutions deprives students of opportunities for development and growth (Aristovnik et al., 2020).

In Bangladesh, COVID-19 has been a major threat due to the high population density (Sarkar et al., 2021). To prioritize public health and safety, the Government of Bangladesh (GoB) has implemented public holidays since March 26, 2020, affecting all institutions (Shammi et al., 2020). In the past, traditional face-to-face classes were conducted, but in the current situation, teachers and students are utilizing virtual technology for educational activities for the first time in the country (Sarkar et al., 2021; Al-Amin et al., 2021). This shift to online teaching and examinations presents significant challenges for both students and educators (Al-Amin et al., 2021). Many students travel from remote areas to attend colleges and universities (Ramij & Sultana, 2020). However, internet costs are high in Bangladesh, and a large portion of the country lacks reliable internet connections (Al-Amin et al., 2021). As a result, students have to rely on expensive internet packages provided by cell phone companies, despite facing financial obstacles (Rahman, 2019). Additionally, poor network coverage often hinders online teaching and learning (Al-Amin et al., 2021).

The impacts of COVID-19 are being felt across populations, with college learners being among the most affected. Uncertainty regarding academic achievement, future careers, and social life in college has caused increased levels of nervousness, depressive moods, psychosomatic problems, lack of self-esteem, substance abuse, and suicidality among students worldwide. As a result, additional services and resources are needed to address the mental and physical health repercussions of the disease. According to Odriozola-González et al. (2020), the continued spread of COVID-19 will have a significant impact on educational institutions, affecting education, social life, and the mental health of students. Studies conducted by Aucejo et al. (2020) have found that students from less privileged backgrounds have been particularly affected by the pandemic, experiencing decreases in family income, limited access to digital resources, and higher costs of internet connectivity, all of which have disrupted their academic lives. Additionally, approximately 1.5 billion students worldwide are currently deprived of basic education due to COVID-19, leading to serious psychological issues. Changes in daily routines, lack of outdoor activity, disrupted sleep patterns, and social isolation have further impacted the psychological well-being of students. Cao et al. (2020) utilized the 7-item GAD-7 (Generalized Anxiety Disorder Scale) as an assessment tool for anxiety disorder, panic disorder, and social phobia. Ye et al. (2020) examined the mediating roles of resilience, coping, and social support in dealing with psychological symptoms. Raju (2020) suggested in his article that policymakers in the education system should adopt innovative teaching techniques to ensure continued education for students and to alleviate anxieties and mental stress during lockdown. It is important to note that the mental health crisis has rapidly emerged across the globe since the COVID-19 pandemic (Czeisler et al., 2020). For students, the increase in depression has coincided with stay-at-home orders, campus closures, and social distancing measures, causing significant disruptions to daily life and altering the way people live, study, and interact (Lee, 2020). According to the findings of Li et al. (2020), college students have been more prone to anxiety, stress, and depression compared to others due to the pandemic. These studies suggest the need to promote and monitor the mental health of young people, especially students, in order to mitigate the negative impact of social isolation (CSTS, 2020; Fessell & Goleman, 2020; Li et al., 2020).

Because schools are designated as essential sources of health, social protection, nutrition, and psychosocial support for children and young adults, students are at a disadvantage in terms of school time as well as social and psychological interactions (Uddin, 2020). The COVID-19 epidemic poses a serious threat to economic security, physical security, and confidence in educational institutions. These threats affect students' emotional and cognitive behavior, leading to an increased use of social media platforms such as Facebook, Messenger, WhatsApp, TikTok, and more, to stay connected with their peers around the world (Ellis, Dumas, & Forbes, 2020). Mamun et al. (2020) emphasize that addiction to social media is associated with unhappiness and poor emotional satisfaction, regardless of the country of residence. Duane et al. (2020) and Wong et al. (2020) suggest that increased internet gaming is one of the main reasons for low mental and physical well-being, as well as the absence of a good night's sleep. Adolescents in Europe and the West are increasingly affected by the epidemic lockdowns and are more likely to experience emotional and physical distress due to the pandemic (Ko & Yen, 2020). Parallelly, the unemployment rate is also rising day by day due to the COVID-19 epidemic situation. Students are frustrated with their future job prospects. Many organizations are unable to carry out their regular activities effectively during the epidemic, resulting in a lack of job opportunities for newcomers.

However, most of the recent literature on the impact of COVID-19 on health and social life has been conducted on the general population, employees, children, and other professionals. Some studies have focused on the mental health of students, while others have examined their social life to some extent. There is limited evidence on the psychological and mental well-being effects of the recent pandemic specifically on students, who are considered a vulnerable group. Although the findings from this research consistently indicate an increase in mental health concerns among students, the contributing factors may not necessarily be applicable to populations in other countries. It is crucial to assess the effects of the current pandemic on the health and social life of students.

17.3 OBJECTIVES OF THE STUDY

Based on the assessment of current literature, the researchers have formulated the following focal objectives of the present study:

- To find out the consequences of the COVID-19 pandemic in students' education life;
- To explore the effects of the COVID-19 pandemic on students' social life;
- To assess the impacts of the COVID-19 pandemic on students' regular education.

17.4 HYPOTHESES OF THE STUDY

➢ **H1:** The closure of schools disrupts students' regular education and increases student violence, having a significant impact on their lives.

- **H2:** During the COVID-19 period, the increase in poverty, unemployment, and social media addiction has had a significant impact on students' lives.

17.5 METHODOLOGY OF THE STUDY

A quantitative method was used to achieve the objectives of this study. Primary data used in the study were collected through a structured questionnaire. A suitable questionnaire was designed, taking into account the objectives of the study, to collect primary data. A survey technique was applied to obtain data from the respondents, using a 5-point Likert scale. A simple random sampling method was used to select the sample unit. The study followed the guidelines of Fishers et al. (1998) to determine the minimum sample size. A total of 804 fully completed questionnaires were used in the study, with 402 from India and 402 from Bangladesh. The Smart PLS 3.2.8 and Statistical Package for Social Science techniques were used to analyze and examine the data. Various statistical tools, such as descriptive analysis, reliability and validity analysis, collinearity statistical analysis, and hypothesis testing, were employed to achieve the objectives and draw the findings of this study.

17.6 DATA ANALYSIS AND INTERPRETATION OF DATA

Figure 17.1 shows the study model by smart PLS (partial least square) structural equation modeling.

17.7 DATA ANALYSIS AND INTERPRETATION

Table 17.1 gives the demographic information of the respondents.

The inputs of Table 17.1 indicate that the majority of the respondents were university and college students. Out of the total 804 respondents, 91.2% were students, 5.5% were teachers, and only 3.4% were guardians. A majority of the respondents, 81.1%, belonged to the age group of 18-24 years. 14.8% of the respondents were between 25 and 34 years old, 3.2% were between 35 and 44 years old, and only 0.9% were 45 years old and older. Among the total respondents, 58.3% were male and 41.7% were female.

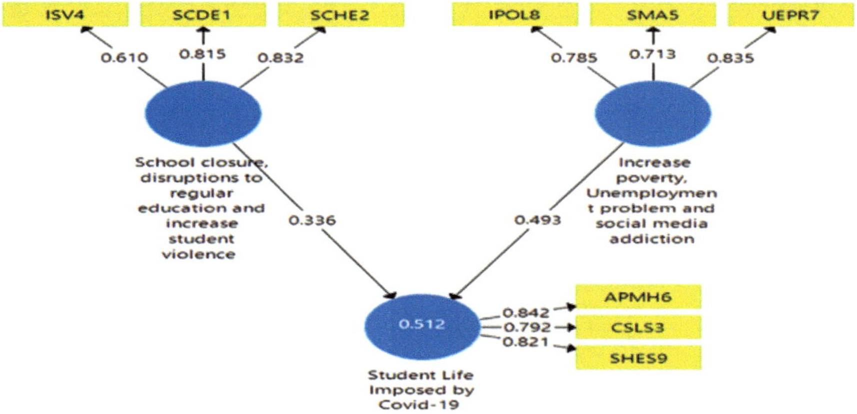

FIGURE 17.1 The study model by smart PLS (partial least square) structural equation modeling.

TABLE 17.1 Demographic Profile of the Respondents

Subject	Subgroup	Frequency	Percentage (%)
Age	18–24 years	652	81.1
	25–34 years	119	14.8
	35–44 years	26	3.2
	45 and older	7	.9
Gender	Male	469	58.3
	Female	335	41.7
Category of respondents	Guardian	27	3.4
	Teacher	44	5.5
	Student	733	91.2

Source: Study survey.

R^2 is a significant statistical tool for assessing a model's analytical ability (Shmueli & Koppius, 2011; Rigdon, 2012). The significant value of R^2 typically ranges from 0 to 1, and a higher R^2 value indicates a better explanatory ability. In this study, the smart PLS technique was used to calculate R^2. Table 17.2 presents the results of the R^2 value, which is 0.512, indicating a moderate explanatory power of the model (Table 17.3).

The metric provided above is the F^2 effect size, which may be considered redundant when compared to the size of the path coefficients. In cases where a dependent construct is explained in the structural model, if there is a difference in the rank order of the constructs' relevance between the size of the path coefficients and the F^2 effect sizes, the researcher may report the F^2 effect size to explain the presence of partial or full mediation (Nitzl et al., 2016). Typically, values higher than 0.02, 0.15, and 0.35 respectively indicate small, medium, and large F^2 effect sizes (Cohen, 1988).

TABLE 17.2 R^2 Value of Smart PLS (Partial Least Square) Structural Equation Modeling

Statement	**R^2**	**Adjusted R^2**
Impediments of student life imposed by COVID-19	0.512	0.511

Source: Calculated by PLS.

TABLE 17.3 F^2 Value of Smart PLS (Partial Least Square) Structural Equation Modeling

Construct	**Impediments of Student Life Imposed by COVID-19 (f Square Value)**	**Comment**
Increase poverty, unemployment problem and social media addiction.	0.388	Large effect
School closure, disruptions to regular education and increase student violence.	0.180	Medium effect

Source: Calculated by PLS.

Table 17.4 shows the results of the measurement model. Generally, factor loadings on a construct were used to measure the indicator reliability of the study. High loadings indicate the associated indicators, which are captured by the construct (Hair et al., 2017). Factor loading values greater than 0.50 were considered to be very significant (Hair et al., 2010). In this study, the factor loadings for all items exceeded the recommended value of 0.50, indicating suitability for analysis. Convergent validity measures the extent to which the items of a scale are theoretically related and correlated. The composite reliability (CR) values of each construct are 0.70 or above, and the average variance extracted (AVE) value of each construct is more than 0.50, which is deemed acceptable. The table shows that all the CR values are above 0.80, ranging from 0.800 to 0.859. The AVE values are above 0.50, ranging from 0.576 to 0.670. Therefore, it can be concluded that the convergent validity has been established (Table 17.5).

TABLE 17.4 The Composite Reliability and Average Variance Extracted Construct

Construct	Item	Loading	Cronbach's Alpha	Composite Reliability	Average Variance Extracted (AVE)
School closure, disruptions to regular education and increase student violence.	ISV4	0.610	0.619	0.800	0.576
	SCDE1	0.815			
	SCHE2	0.832			
Increase poverty, unemployment problem and social media addiction.	SMA5	0.713	0.674	0.822	0.608
	UEPR7	0.835			
	IPOL8	0.785			
Impediments of student life imposed by COVID-19	APMH6	0.842	0.754	0.859	0.670
	CSLS3	0.792			
	SHES9	0.821			

Source: Calculated by PLS.

TABLE 17.5 The Discriminant Validity (Fornell-Larcker Criterion)

Construct	1	2	3
Increase poverty, unemployment problem and social media addiction.	0.780	–	–
School closure, disruptions to regular education and increase student violence.	0.472	0.759	–
Impediments of student life imposed by COVID-19.	0.652	0.568	0.819

Source: Calculated by PLS.

Hair et al. (2017) suggested that discriminant validity is used to confirm a reflective concept that has strong associations with its own indicators in PLS path modeling. Henseler et al. (2015) recommend using the HTMT principle to assess discriminant validity. In the present case, the HTMT calculation value is below 0.90, indicating that discriminant validity has been confirmed between the two constructs under consideration. The results clearly indicate that discriminant validity is well established.

As predicted, the findings in Table 17.6 confirmed that the impediments of student life imposed by the COVID-19 pandemic situation include school closures, disruptions to regular education, and increased student violence ($\beta = 0.339$, $T = 8.605$, $p < 0.000$). Therefore, H1 is strongly supported. Additionally, the COVID-19 pandemic situation also leads to increased poverty, unemployment problems, and social media addiction (β= 0.492, T = 11.135, p< 0.000). Hence, H2 is strongly supported.

TABLE 17.6 The Path Coefficients and T-Statistics

Hypothesized Path	Standardized Beta	T Statistics	P Values	Comment
H1: School closure, disruptions to regular education and increase student violence → impediments of student life imposed by COVID-19.	0.339	8.605	0.000	Supported
H2: Increase poverty, unemployment problem and social media addiction → impediments of student life imposed by COVID-19.	0.492	11.135	0.000	Supported

Source: Calculated by PLS.

17.8 CONCLUSION

The COVID-19 outbreak has had adverse effects on students' education, health, economy, and societal relationships. Measures such as the closure of educational institutions, social distancing, quarantine, and others have significantly impacted the teaching and learning experiences of students during the pandemic. The forced shutdown of regular social communication patterns has disrupted economic and social activities. This study focuses on the effects of COVID-19 on students' education and social life in India and Bangladesh.

The study found that disruptions in teaching and learning, difficulties in social relationships, mental and physical health issues, and financial crises resulting from the virus outbreak have created numerous challenges for students' academic performance and overall behavior. Additionally, the shift from face-to-face education to online learning has caused significant stress, disappointment, and anxiety among students. The findings of this study aim to provide constructive feedback to teachers, educators, and government officials involved in the education sector in India and Bangladesh. It is expected that support from friends, family, communities, and the government will be crucial in mitigating the social, physical, mental, and educational disruptions faced by students.

KEYWORDS

- **Bangladesh**
- **COVID-19 pandemic**
- **data analysis**
- **education**
- **educational disruptions**
- **India**
- **social media addiction**
- **student**

REFERENCES

Ainscow, M., & Haile-Giorgis, M., (1999). Educational arrangements for children categorized as having special needs in Central and Eastern Europe. *European Journal of Special Needs Education, 14*(2), 103–121.

Ainscow, M., (2005). Developing inclusive education systems: What are the levers for change? *Journal of Educational Change, 6*(2), 109–124.

Al-Amin, M., Al Zubayer, A., Deb, B., & Hasan, M., (2021). Status of tertiary level online class in Bangladesh: Students' response on preparedness, participation, and classroom activities. *Heliyon, 7*(1), e05943.

Aristovnik, A., Kerzic, D., Ravselj, D., Tomazevic, N., & Umek, L., (2020). Impacts of the COVID-19 pandemic on life of higher education students: A global perspective. *Sustainability, 12*(20), 8438.

Aucejo, E. M., French, J., Araya, M. P. U., & Zafar, B., (2020). The impact of COVID-19 on student experiences and expectations: Evidence from a survey. *Journal of Public Economics, 191*, 104271.

Błachnio, A., Przepiorka, A., Benvenuti, M., Mazzoni, E., & Seidman, G., (2019). Relations between Facebook intrusion, internet addiction, life satisfaction, and self-esteem: A study in Italy and the USA. *International Journal of Mental Health and Addiction, 17*(4), 793–805. https://doi. org/10.1007/s11469-018-0038-y.

Cao, W., Fang, Z., Hou, G., Han, M., Xu, X., Dong, J., & Zheng, J., (2020). The psychological impact of the COVID-19 epidemic on college students in China. *Psychiatry Research, 287*, 112934.

CDC, (2020). *Coronavirus Disease 2019 (COVID-19) – FAQs for Administrators*. Centers for Disease Control and Prevention.

Cohen, J., (1988). *Statistical Power Analysis for the Behavioral Sciences.* Lawrence Erlbaum Associates.

Czeisler, M. E., Lane, R. I., Petrosky, E., Wiley, J. F., Christensen, A., Njai, R., & Rajaratnam, S. M., (2020). Mental health, substance use, and suicidal ideation during the COVID-19 pandemic—United States. *Morbidity and Mortality Weekly Report*, *69*(32), 1049.

Dhawan, S., (2020). Online learning: A panacea in the time of COVID-19 crisis. *Journal of Educational Technology Systems*, *49*(1), 5–22.

Duan, L., Shao, X., Wang, Y., Huang, Y., Miao, J., Yang, X., & Zhu, G., (2020). An investigation of mental health status of children and adolescents in China during the outbreak of COVID-19. *Journal of Affective Disorders*, *275*, 112–118. https://doi.org/10.1016/j.jad.2020.06.029.

Ellis, W. E., Dumas, T. M., & Forbes, L. M., (2020). Physically isolated but socially connected: Psychological adjustment and stress among adolescents during the initial COVID-19 crisis. *Canadian Journal of Behavioral Science*, *52*(3), 177–187. https://doi.org/10.1037/cbs0000215.

Fessell, D., & Goleman, D., (2020). *How Healthcare Personnel Can Take Care of Themselves.* US: HBR.

Fisher, A. A., Laing, J. E., Stoeckel, J. E., & Townsend, J. W., (1998). *Handbook for Family Planning Operations Research Design.* Population Council, New York.

Hair, J. F., Black, W. C., Babin, B. J., & Anderson, R. E., (2010). *Multivariate Data Analysis.* Upper Saddle River, NJ: Prentice-Hall.

Hair, J. F., Hult, G. T. M., Ringle, C., & Sarstedt, M., (2017). *A Primer on Partial Least Squares Structural Equation Modeling (PLS-SEM)* (2nd edn.). Sage, Thousand Oaks, London.

Harsha, R., & Bai, T., (2020). Covid-19 lockdown-challenges to higher education. *Cape Comorin*, *2*(4), 26–28.

Henseler, J., Ringle, C. M., & Sarstedt, M., (2015). A new criterion for assessing discriminant validity in variance-based structural equation modeling. *Journal of the Academy of Marketing Science*, *43*(1), 115–135.

Hulin, C., Netemeyer, R., & Cudeck, R., (2001). Can a reliability coefficient be too high? *Journal of Consumer Psychology*, *10*, 55–58.

Institute of Epidemiology, Disease Control and Research (IEDCR), (2020). *Covid-19 Vital Statistics*. IEDCR. https://iedcr.gov.bd (accessed on 11 January 2023).

Joshi, A., Vinay, M., & Bhaskar, P., (2020). Impact of coronavirus pandemic on the Indian education sector: Perspectives of teachers on online teaching and assessments. *Interactive Technology and Smart Education*, *18*(2), 205–226.

Ko, C. H., & Yen, J. Y., (2020). Impact of COVID-19 on gaming disorder: Monitoring and prevention. *Journal of Behavioral Addictions*, *9*(2), 187–189. https://doi.org/10.1556/2006.2020.00040.

Kumar, D., Adhikari, P. K., & Kumar, S., (2020). COVID-19: Its instant impacts upon the agrarian economy of Bangladesh. *J. Bus. Econ. Manag.*, *8*(8), 193–199.

Kumar, D., Bagchi, S., & Ray, S., (2021). Covid-19 and risk involvement in the banking sector of Bangladesh: A study. *International Journal of Research – GRANTHAALAYAH*, *6*, 158–167.

Lee, J., (2020). Mental health effects of school closures during COVID-19. *The Lancet Child & Adolescent Health*, *4*(6), 421.

Li, L. Z., & Wang, S., (2020). Prevalence and predictors of general psychiatric disorders and loneliness during COVID-19 in the United Kingdom. *Psychiatry Research, 291*, 113267.

Mamun, M. A., Hossain, M. S., Moonajilin, M. S., Masud, M. T., Misti, J. M., & Griffiths, M. D., (2020). Does loneliness, self-esteem, and psychological distress correlate with problematic internet use? A Bangladeshi survey study. *Asia-Pacific Psychiatry, 12*(2). https://doi.org/10.1111/appy.12386.

Murphy, M. P., (2020). COVID-19 and emergency eLearning: Consequences of the securitization of higher education for post-pandemic pedagogy. *Contemporary Security Policy, 41*(3), 492–505.

Nitzl, C., Roldán, J. L., & Cepeda, C. G., (2016). Mediation analysis in partial least squares path modeling: Helping researchers discuss more sophisticated models, *Industrial Management and Data Systems, 119*(9), 1849–1864.

Ocriozola-González, P., Planchuelo-Gómez, A., Irurtia, M. J., & De Luis-García, R., (2020). Psychological effects of the COVID-19 outbreak and lockdown among students and workers of a Spanish university. *Psychiatry Research, 290*, 113108.

OECD, (2020). *Development A Framework to Guide an Education Response to the COVID-19 Pandemic of 2020*. Retrieved from: https://read.oecd-ilibrary.org/view/?ref=126_126988-t63lxosohs&title=A-framework-to-guide-an-education-response-to-the-Covid-19-Pandemic-of-2020 (accessed on 11 January 2023).

OECD, (2020). *Learning Remotely When Schools Close: How Well are Students and Schools Prepared?* Insights from PISA. Retrieved from: https://read.oecd-ilibrary.org/view/?ref=127_127063-iiwm328658&title=Learning-remotely-when-schools-close (accessed on 11 January 2023).

Pokhrel, S., & Chhetri, R., (2021). A literature review on impact of COVID-19 pandemic on teaching and learning. *Higher Education for the Future, 8*(1), 133–141.

Rahman, A., (2019). E-governance in Bangladesh. In: Farazmand, A., (ed.), *Global Encyclopedia of Public Administration, Public Policy, and Governance* (pp. 1–8). Cham: Springer. doi: 10.1007/978-3-319-31816-5_3227-1.

Ramij, M. G., & Sultana, A., (2020). *Preparedness of Online Classes in Developing Countries Amid COVID-19 Outbreak: A Perspective from Bangladesh*. doi: 10.2139/ssrn.3638718.

Rigdon, E. E., (2012). Rethinking partial least squares path modeling: In praise of simple methods. *Long Range Planning, 45*(5, 6), 341–358.

Rodricks, Z., (2021). *Covid-19 and its Impact on Education System*. The Times of India. Retrieved from: https://timesofindia.indiatimes.com/readersblog/zita-janice/covid-19-and-its-impact-on-education-system-35076/ (accessed on 11 January 2023).

Roy, N. C. (2022). Impact of COVID-19 on the migrant worker livelihood: Study in the Jalpaiguri Region, India. *International Journal of Services, Economics and Management.*

Sarkar, S. S., Das, P., Rahman, M. M., & Zobaer, M. S., (2021). Perceptions of public university students towards online classes during COVID-19 pandemic in Bangladesh. In: *Frontiers in Education* (Vol. 6, p. 265). Frontiers.

Shammi, M., Bodrud-Doza, M., Islam, A. R. M. T., & Rahman, M. M., (2020). COVID-19 pandemic, socioeconomic crisis, and human stress in resource limited settings: A case from Bangladesh. *Heliyon, 6*(5), e04063. doi: 10.1016/j.heliyon.2020.e04063.

Shmueli, G., & Koppius, O. R., (2011). Predictive analytics in information systems research. *MIS Quarterly, 35*(3), 553–572.

Skulmowski, A., & Rey, G. D., (2020). COVID-19 as an accelerator for digitalization at a German university: Establishing hybrid campuses in times of crisis. *Human Behavior and Emerging Technologies, 2*(3), 212–216.

Suresh, A., (2021). *Impact of Covid-19 on School Education in India.* Times of India. Retrieved from: https://timesofindia.indiatimes.com/readersblog/theenchantedpen/impact-of-covid-19-on-school-education-in-india-32475/ (accessed on 11 January 2023).

The Hindu, (2020). *CBSE Board Exams to be Held in Schools Where Students Enrolled, Result by July-End.* Retrieved from: https://www.thehindu.com/news/national/cbse-board-exams-to-be-held-in-schools-where-students-enrolled-result-by-july-end/article 31629821.ece (accessed on 11 January 2023).

The Telegram, (2020). *Coronavirus Deprives Nearly 300 Million Students of Their Schooling: UNESCO.* Retrieved from: https://www.saltwire.com/newfoundland-labrador/news/world/coronavirus-deprives-nearly-300-million-students-of-their-schooling-unesco-419714/ (accessed on 11 January 2023).

Uddin, M., & Uddin, B., (2021). The impact of COVID-19 on students' mental health. *Journal of Social, Humanity, and Education, 1*(3), 185–196.

Uddin, M., (2020). *Effects of the Pandemic on the Education Sector in Bangladesh.* The FinancialExpress. https://thefinancialexpress.com.bd/views/effects-of-the-pandemic-on-the-education-sector-in-bangladesh-1592061447 (accessed on 11 January 2023).

UNESCO, (2020). *290 Million Students Out of School Due to COVID-19: UNESCO Releases First Global Numbers and Mobilizes Response.* Retrieve from: https://en.unesco.org/news/290-million-students-out-school-due-covid-19-unesco-releases-first-global-numbers-and-mobilizes (accessed on 11 January 2023).

UNESCO, (2020). *Adverse Consequences of School Closures.* Retrieved from: https://en.unesco.org/covid19/educationresponse/consequences (accessed on 11 January 2023).

UNESCO, (2021). *School Closure Caused by Coronavirus (COVID-19).* https://www.unesco.org/en/covid-19/education-response

UNICEF, (2020). *COVID-19: Schools for More Than 168 Million Children Globally Have Been Completely Closed for Almost a Full Year, Says UNICEF.* Retrieved from: https://www.unicef.org/india/press-releases/covid-19-schools-more-168-million-children-globally-have-been-completely-closed (accessed on 11 January 2023).

UNICEF, (2021). *The Future of 37 Million Children in Bangladesh is at Risk with Their Education Severely Affected by the COVID-19 Pandemic.* https://www.unicef.org/bangladesh/en/press-releases/future-37-million-children-bangladesh-risk-their-education-severely-affected-covid (accessed on 11 January 2023).

United Nations, (2020). *Opening Remarks at Virtual Press Encounter to Launch the Report on the Socio-Economic Impacts of COVID-19.* Retrieved from: www.un.org/sg/en/content/sg/speeches/2020-03-31/remarks-launch-of-report-the-socio-economic-impacts-of-covid-19 (accessed on 11 January 2023).

Vuorikari, R., Velicu, A., Chaudron, S., Cachia, R., & Di Gioia, R., (2020). *How Families Handled Emergency Remote Schooling During the COVID-19 Lockdown in Spring 2020: Summary of Key Findings from Families with Children in 11 European Countries.* JRC Science for Policy Report. European Commission. Available from: EU Bookshop.

Wang, C., Pan, R., Wan, X., Tan, Y., Xu, L., Ho, C. S., & Ho, R. C., (2020). Immediate psychological responses and associated factors during the initial stage of the 2019 coronavirus disease (COVID-19) epidemic among the general population in China. *International Journal of Environmental Research and Public Health, 17*(5), 1729.

WHO, (2020). *WHO Director-General's Opening Remarks at the Media Briefing on COVID-19*. Retrieve from: www.who.int/director-general/speeches/detail/who-director-general-sopening-remarks-at-the-media-briefing-on-covid-19---11-march-2020 (accessed on 11 January 2023).

Wong, H. Y., Mo, H. Y., Potenza, M. N., Chan, M. N. M., Lau, W. M., Chui, T. K., & Lin, C. Y., (2020). Relationships between severity of internet gaming disorder, severity of problematic social media use, sleep quality and psychological distress. *International Journal of Environmental Research and Public Health, 17*(6), 1879. https://doi.org/10.3390/ijerph17061879.

Yeh, Z., Yang, X., Zeng, C., Wang, Y., Shen, Z., Li, X., & Lin, D., (2020). Resilience, social support, and coping as mediators between COVID-19 related stressful experiences and acute stress disorder among college students in China. *Applied Psychology: Health and Well-Being, 12*(4), 1074–1094.

CHAPTER 18

Effectiveness of Social Media Marketing as a Sustainable Futuristic Tool for Consumer Decision-Making Process

AHASANUL HAQUE[1] and KHUSHBU AGARWAL[2]

[1]*Department of Business Administration, International Islamic University Malaysia, Kuala Lumpur, Malaysia*

[2]*Pacific Institute of Management, Pacific University, Udaipur, Rajasthan, India*

ABSTRACT

The most recent development in Industry 4.0 and the marketing sector is social media marketing (SMM). Making impactful conversations is the cornerstone of success for any media (Agarwal, 2013). The latest sustainable marketing strategy is social media marketing, which is built on the widely accepted idea of word-of-mouth marketing (WOM). India is currently the seventh-largest internet market in the world, with a growth rate of 11.2%. India now has 60.5 million active social media users, up from 38 million in November 2011 (Agnihotri, Kothandaraman, Kashyap, & Singh, 2012). In a nutshell, social media websites in India see annual growth rates of around 100%. Despite the tremendous growth of social media in India over the past few years, there are still many untapped potentials (Amerian & Pisarava, 2012). This research aims to raise awareness among marketers of Industry 4.0 and Industry 5.0 worldwide about the power of social media in spreading sustainable

Sustainable Strategic Business Infrastructure Development and Contemporary Digital Practices in Industry 5.0. Indrajit Ghosal, Bikram Prasad, Nilanjan Ray, & Lalit Kumar Joshi (Eds.)

environmental messages at large. The impact of sustainable social media marketing strategies on consumer behavior is the main focus of the current research work. A sample of 358 social media users was chosen to achieve the goal. According to the report, the majority of consumers are most aware of social media marketing and product photographs and videos. Findings indicated that making sustainable buying decisions on social media is greatly influenced by how easily one can find information. Customers were asked to indicate their satisfaction with different factors of social media, and they are highly satisfied with the ease of searching for information on social networks. Poor response systems, physical unavailability of products, improper replacement facilities, and lack of reliability of information were considered major barriers by customers in stimulating or satisfying their purchase decisions.

18.1 INTRODUCTION

18.1.1 SOCIAL MEDIA MARKETING

Social media is currently a buzzword and in high demand. Additionally, it offers organizations a marketing opportunity that bypasses the conventional middleman and puts them in direct contact with clients. Businesses were unsure about social media a year ago. Social media marketing is becoming a permanent fixture, and businesses are quickly embracing it. Social networking is the next marketing trend, just as email and websites were the first to empower businesses.

Social media marketing includes advertising on blogs, social networks, online forums, and other platforms. It's the latest marketing trend. India is likely one of the early adopters of social media marketing. Nowadays, businesses are focusing on connecting with their customers through internet channels rather than social causes.

The rapid growth of this phenomenon is astonishing, and the explosion of social media is even more remarkable. Trust and kindness are the foundation of social networking, and these fundamental principles must be upheld while promoting on social media. Global businesses have recognized social media marketing as an effective marketing channel and have utilized it to enhance their sustainable advertising campaigns with innovative approaches.

18.2 RESEARCH METHODOLOGY

18.2.1 OBJECTIVES OF THE STUDY

- To assess the role of sustainable Social Media marketing in decision making process.
- To assess the effectiveness of social media marketing as a sustainable tool in comparison to other traditional promotional activities.
- To examine the obstacles in social media marketing that influence customers' purchasing decisions and satisfaction levels.

➢ **Hypothesis:** "There is no relationship between social media marketing and customers' purchasing intention."

18.2.2 SAMPLING

The sample for the study has been selected using the stratified purposive sampling technique to represent Rajasthan. To ensure a representative sample of Rajasthan, it has been divided into 5 strata, which are the five major cities of Rajasthan: Jaipur, Jodhpur, Ajmer, Kota, and Udaipur.

These sample units have been purposively selected from social media users only. To make the samples representative, a total of 348 social media users' data have been gathered from all five selected cities/districts of Rajasthan in the following proportions: Jaipur (67), Jodhpur (66), Udaipur (104), Kota (56), and Ajmer (55).

18.2.3 DATA COLLECTION

The study was conducted using primary, secondary, and other qualitative inputs that directly influence social media marketing practices. The research relied on field survey techniques, such as questionnaires, interviews, and observations, as well as published and unpublished reports, records, journals, periodicals, newspapers, and magazines to collect primary and secondary data. The primary information was gathered from social media users in Rajasthan. The users were surveyed through a well-structured questionnaire.

18.2.4 RELIABILITY OF DATA COLLECTION INSTRUMENT

The Cronbach's alpha coefficient is frequently used by researchers to determine scale dependability. The scale's internal consistency is shown by the Cronbach's alpha coefficient. Cronbach's alpha can be used as a suitable test of scale dependability when it is more than 0.60. Reliability results for the questionnaire are presented in Tables 18.1 and 18.2.

TABLE 18.1 Case Processing Summary

		N	Percentage (%)
Cases	Valid	348	100.0
	Excluded[a]	0	0.0
	Total	348	100.0

[a]List wise deletion based on all variables in the procedure.

TABLE 18.2 Reliability Statistics

Number of Items	Cronbach's Alpha
52	0.847

Table 18.2 shows that the value of Cronbach's alpha is 0.847. As suggested by previous researchers, an acceptable level of reliability for a psychometric test starts from 0.60. This indicates a good internal consistency of the items in the scale.

18.3 DATA ANALYSIS AND INTERPRETATION

18.3.1 ROLE OF SUSTAINABLE SOCIAL MEDIA MARKETING IN DECISION MAKING PROCESS

This section of the chapter will discuss the decision-making process on social media. We asked respondents to rate their level of agreement with specific statements. The scale for indicating agreement ranged from 1 to 5, with 5 representing strong agreement and 1 representing strong disagreement.

Additionally, the analytical part uses the following criteria:

- The score among 1.00–1.80 means strongly disagree.

- The score among 1.81–2.60 means disagree.
- The score among 2.61–3.40 means neutral.
- The score among 3.41–4.20 means agree.
- The score among 4.21–5.00 means strongly agree.

As per the analysis shown in Table 18.3, out of the total respondents (N=348), the majority of the respondents (N=165, Percentage=47.41%) neither agree nor disagree with the statement that social media makes the decision-making process more complex. This is followed by N=89, i.e., 25.57% of respondents who disagree with it.

21.84% (N=76) of respondents agree with the fact that social media makes the decision process more complex, whereas only 13 strongly disagree and 5 strongly agree with it.

The overall mean score of 2.917 indicates that the majority of consumers neither agree nor disagree with the statement, i.e., social media makes the decision process more complex (Figure 18.1).

TABLE 18.3 Role of Sustainable Social Media Marketing in Decision Making Process

Level	N	Percentage (%)	Cumulative Percentage
Strongly agree	5	1.44%	1.44%
Agree	76	21.84%	23.28%
Neither agree nor disagree	165	47.41%	70.69%
Disagree	89	25.57%	96.26%
Strongly disagree	13	3.74%	100.00%
Total	348	100%	–
Mean score	2.917		
Standard deviation	0.822		
Result	Neither agree nor disagree		

18.3.2 SOCIAL MEDIA MARKETING VS. TRADITIONAL PROMOTION ACTIVITIES

This section of the chapter will discuss the comparison between social media marketing and traditional marketing tools. We requested the respondents to indicate the level of effectiveness of social media marketing in comparison to traditional marketing activities. The effectiveness is measured on a scale of 1 to 5 for different attributes of social media (where 5 represents "Much more effective" and 1 represents "Much less effective").

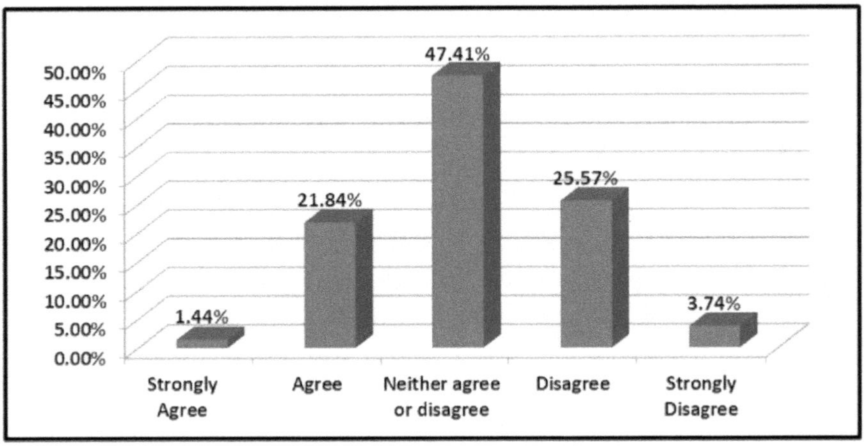

FIGURE 18.1 Role of social media marketing in decision making process.

In addition, following criteria is used for analysis part:

- The score among 1.00–1.80 means much less effective.
- The score among 1.81–2.60 means less effective.
- The score among 2.61–3.40 means as effective.
- The score among 3.41–4.20 means more effective.
- The score among 4.21–5.00 means much more effective.

According to the analysis presented in Table 18.4, out of the total respondents (N=348), the majority of the respondents (N=117, Percentage=33.62%) believe that social media marketing is as effective as traditional marketing activities. This is followed by 109 respondents who consider it to be more effective than traditional marketing.

The overall mean score of 3.247 projects that customers consider social media marketing as effective as traditional marketing activities (Figure 18.2).

18.4 BARRIERS OF SMM THAT STIMULATE CUSTOMERS' PURCHASE DECISION/SATISFACTION LEVEL

This section of the chapter will discuss barriers that influence customers' purchase decisions and satisfaction levels. It will explore whether the disadvantages of social media act as barriers and, if so, to what extent. Respondents were asked to indicate the level of barrier for different critical factors of social media that affect their purchase decisions and satisfaction. The degree of barrier was rated on a scale of 1 to 4 for various attributes of social media, with 4 indicating an extreme barrier and 1 indicating no barrier.

TABLE 18.4 Effectiveness of Social Media Marketing as a Sustainable Tool in Comparison to Other Traditional Promotion Activities

SL No.	Level	N	Percentage (%)	Cumulative Percentage
1.	Much less effective	12	3.45%	3.45%
2.	Less effective	73	20.98%	24.43%
3.	As effective	117	33.62%	58.05%
4.	More effective	109	31.32%	89.37%
5.	Much more effective	37	10.63%	100.00%
	Total	348	100.00%	–
	Mean	3.247		
	SD	1.014		
	Result	As effective		

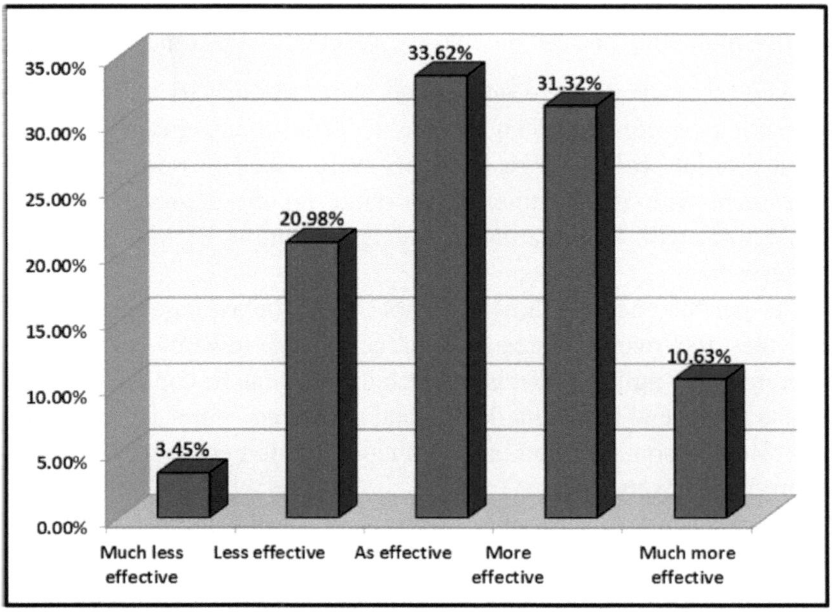

FIGURE 18.2 Effectiveness of social media marketing as a sustainable tool in comparison to other traditional promotion activities.

In addition, following criteria is used for analysis part:

- The score among 1.00–1.75 means not a barrier.
- The score among 1.75–2.50 means somewhat barrier.
- The score among 2.50–3.25 means moderate barrier.
- The score among 3.25–4.00 means Extreme barrier.

Following barriers of SMM have presented in further pages(Table 18.5 and Figure 18.3):

- Poor response system;
- Lack of reliability of the information;
- Advance payment;
- Abundance of information;
- Physical unavailability of product;
- Lack of personal attention;
- Improper delivery system;
- Improper replacement facility;
- Irritating;
- Lack of privacy.

➢ **Hypothesis:** "There is no relationship between social media marketing and the purchasing intention of customers."

The respondents were questioned about the impact of social media marketing on consumers' intentions to acquire any goods or services. On a 5-point scale, 348 respondents were asked to rate their level of agreement with the statement. The range for the degree of agreement was 1 to 5, with 5 indicating strong agreement and 1 indicating strong disagreement.

As per the analysis shown in Table 18.6, the average score of 3.647 indicates the overall agreement of customers towards the statement. Therefore, the null hypothesis is rejected, and it can be concluded from the results that social media marketing and customers' purchase intention are associated in some manner, as it influences customers to make purchases of products or services.

After discussing customers' feelings and satisfaction levels towards social media marketing parameters and various barriers of social media marketing, we will now proceed to the final section, which is the conclusions and suggestions.

TABLE 18.5 Barriers of SMM That Stimulate Customers' Purchase Decision/Satisfaction Level

Sl. No.	Factors of Social Media Marketing	Extreme Barrier	Moderate Barrier	Somewhat Barrier	Not a Barrier	Total	Mean Score	Standard Deviation	Result
1.	Poor response system	155	123	26	44	348	3.118	1.007	Moderate barrier
2.	Lack of reliability of the information	109	135	71	33	348	2.92	0.945	Moderate barrier
3.	Advance payment	73	160	94	21	348	2.819	0.831	Moderate barrier
4.	Abundance of information	61	150	79	58	348	2.615	0.961	Moderate barrier
5.	Physical unavailability of product	127	123	64	34	348	2.986	0.971	Moderate barrier
6.	Lack of personal attention	80	160	54	54	348	2.764	0.976	Moderate barrier
7.	Improper delivery system	122	113	65	48	348	2.888	1.039	Moderate barrier
8.	Improper replacement facility	112	135	81	20	348	2.974	0.887	Moderate barrier
9.	Irritating	67	101	108	72	348	2.468	1.025	Somewhat barrier
10.	Lack of privacy	108	135	59	46	348	2.876	0.998	Moderate barrier

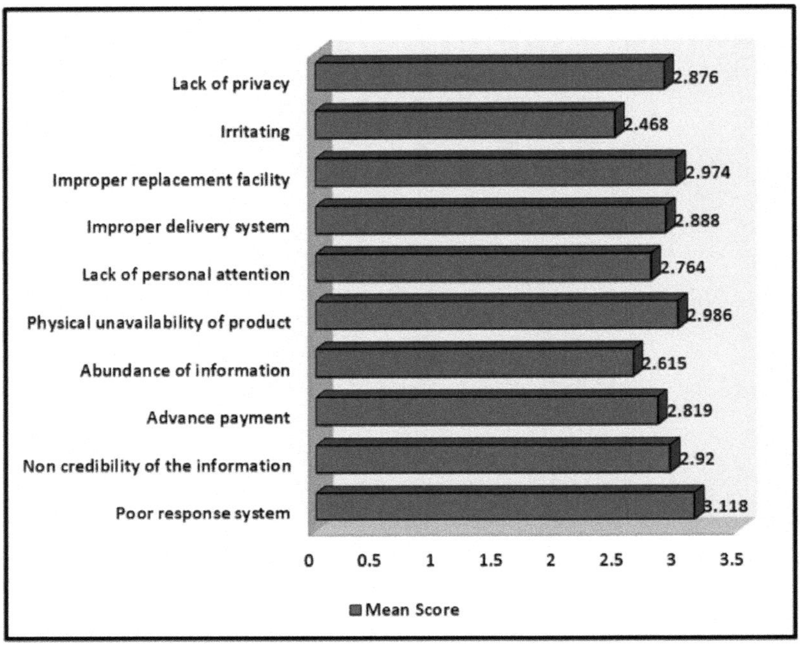

FIGURE 18.3 Barriers of SMM that stimulate customers' purchase decision/satisfaction level.

TABLE 18.6 Role of Social Media Marketing in Purchase of Any Product/Service

Level	N	Percentage (%)	Cumulative Percentage
Strongly agree	54	15.52%	15.52%
Agree	150	43.10%	58.62%
Neither agree nor disagree	114	32.76%	91.38%
Disagree	27	7.76%	99.14%
Strongly disagree	3	0.86%	100.00%
Total	348	100%	
Mean	3.647		
SD	0.865		
Result	Agree		

18.5 CONCLUSION AND RECOMMENDATION

Social networking offers numerous benefits for businesses. Firms must consider their goals for the year and how social media aligns with those goals in order to develop a strategy for its use. One advantage of a social media strategy is the ability to adapt the available tools to meet specific needs. Businesses can choose to focus their efforts on locations that provide the best return on investment, while taking a "wait and see" approach to the rest.

Overall, it is evident that social media marketing attracts customers, and to remain competitive and sustainable in the market, continuous development of social media marketing strategies is required. This study aims to examine the impact of social media marketing practices as a sustainable tool on customer behavior in Rajasthan.

KEYWORDS

- **consumer decision making process**
- **data analysis**
- **data collection**
- **Industry 4.0**
- **satisfaction level**
- **social media marketing**
- **sustainable social media marketing**
- **sustainable tool**

REFERENCES

Agarwal, T., (2013). *Social Media Advertising*. Paper presented at the International Conference on Managing change in Business and Economy, Udaipur, India.

Agnihotri, R., Kothandaraman, P., Kashyap, R., & Singh, R., (2012). Bringing "social" into sales: The impact of salespeople's social media use on service behaviors and value creation. *Journal of Personal Selling & Sales Management, XXXII*(3), 333–348.

Amerian, I., & Pisarava, N., (2012). *Integrating Social Media Into the Marketing Communication Strategy.* Master thesis, Jönköping University, Sweden.

Bashar, A., Ahmad, I., & Wasiq, M., (2012). Effectiveness of social media as a marketing tool: An empirical study. *International Journal of Marketing, Financial Services & Management Research, 1*(11), 88–99.

Bhakuni, P., & Aronkar, P., (2012). Effect of social media advertising on purchase intentions of students: an empirical study conducted in Gwalior city. *International Journal of AppliedServices Marketing Perspectives, 1*(1), 73–79.

Bhattarcharyya, D., (2005). *Research Methodology*. Excel Books, New Delhi.

D'Silva, B., Bhuptani, R., Menon, S., & D'Silva, S., (2011). *Influence of Social Media Marketing on Brand Choice Behavior Among Youth in India: An Empirical Study* (pp. 756–763). Presented in International Conference on Technology and Business Management.

Dash, A. K., (2011). Use of online social net-working sites by college students and its implication for marketing: A case study in Tripura. *Indian Journal of Marketing*, 68–76.

Hollensen, S., (2007). *Global Marketing: A Marketing Responsive Approach.* Prentice Hall.

Keppel, G., (2016). *Design and Analysis. A Researcher's Handbook.* Prentice Hall, New Jersey.

Kothari, C. R., (1997). *Research Methodology–Methods and Techniques* (p. 18). Wishwa Prakashan, New Delhi.

Michael, V. P., (1985). *Research Methodology in Management* (p. 107). Himalayan Publishing house, Bombay.

Vij, S., & Sharma, J., (2013). *An Empirical Study on Social Media Behavior of Consumers and Social Media Marketing Practices of Marketers* (pp. 1–19). Presented paper in 5thIIMA Conference on Marketing in Emerging Economies.

Yadav, N., (2012). Social networking sites-a new vehicle for advertising. *MIMT Journal of IT & Management Research, 2*(1), 38–48.

CHAPTER 19

Framework for Adopting a Data Analytics Strategy in the Fourth Industrial Revolution as a Stepping Stone to the Fifth Industrial Revolution

GARIKAYI EMMANUEL GONDO,[1] LORNA B. HOLTMAN,[2] and RAVINDER RENA[3]

[1]*UNISA School of Business Leadership (SBL), University of South Africa (UNISA), Midrand, South Africa*

[2]*Honorary Research Fellow, International Education and Partnerships, Durban University of Technology, ML Sultan Campus, Durban, Republic of South Africa*

[3]*DUT Business School, Faculty of Management Sciences, Durban University of Technology, ML Sultan Campus, Durban, Republic of South Africa*

ABSTRACT

The fourth industrial revolution calls for companies to be creative and change the current way of doing business. This chapter explores Industry 4.0, digital citizenship, and advanced connectivity technologies. The objective of this chapter was to explore what managers need to do to adapt to the new normal of Industry 4.0, where there is now a need to have smart factories by configuring machines and gadgets to supervise humans and report to managers through cyber-physical systems (CPS). The COVID-19

Sustainable Strategic Business Infrastructure Development and Contemporary Digital Practices in Industry 5.0. Indrajit Ghosal, Bikram Prasad, Nilanjan Ray, & Lalit Kumar Joshi (Eds.)

pandemic sent panic waves to most managers across different sectors as they were forced to stay at home and interact with digital tools to monitor and manage their workplaces. This accelerated the movement into the fifth industrial revolution. The new environment requires managers to embrace emerging technologies and big data opportunities and then change the way of doing business. This triggered us to carry out this study. We explored the new requirements, then identified new opportunities for managers in the big data environment to merge the two. As part of this study, we interviewed senior managers and executives. The results revealed that there is a need for a paradigm shift in the current ways of doing business. We then developed a framework to accelerate the adoption of new technologies to improve data analytics.

19.1 INTRODUCTION AND BACKGROUND

Industry evolves over time, and it relies on dominant drivers at any given period. No operational model lasts forever. The first industrial revolution, known as Industry 1.0, was driven by water and steam power until the late 18th century. The second revolution, driven by electrical power, began in the early 20th century and led to mass production. The third revolution, spanning from the 1970s to the 20th century, was driven by electronics and information technology. This brought us to the current fourth industrial revolution, known as Industry 4.0, which is driven by Cyber-Physical Systems (CPS). The term Industry 4.0 was first mentioned at an industrial fair in Hanover, Germany, in 2011 (Khan, Wu, Xu, & Dou, 2017). It was coined as a strategic roadmap to promote digitalization in the manufacturing sector (Santos, Mehrsai, Barros, Araujo, & Ares, 2017). However, these technologies have advanced at a faster rate than expected, leaving managers behind.

Managers have been lining up to digitize processes, but at the same time, they are not very clear on the role of big data in this new environment. They were responding slowly to the demands of Industry 4.0 without a clear strategy to digitalize their operations. The COVID-19 outbreak put everyone into panic mode, calling for people to conduct business remotely. Indeed, digitalization proved to be the way to go as meetings were now being conducted remotely, calling for a paradigm shift in the current ways of doing business. Managers were forced to train workers to work from home while trying to maintain the same level of performance (Al-Habaibeh, Watkins, Waried, & Javareshk, 2021).

The crises sent shockwaves to managers, resulting in the accelerated adoption of digital technologies. However, very few managers now believe they are ready to compete as digital leaders (Snyder & Barnakova, 2020). Managers now have to rely on information generated by interconnected digital devices in this new fourth industrial revolution to operate efficiently. Industry 4.0 is expected to solve the long-standing problem of information failures. Wilen-Sky (1967) extensively discusses this problem in the notion of "information failures," which occur when accurate information is distorted by supervisors, intentionally or unintentionally, as it is transmitted to decision-makers through behaviors referred to as sharpening, leveling, and gatekeeping to avoid answering numerous questions and being exposed (O'Reilly & Roberts, 1974). In this chapter, we present a framework for adopting a big data strategy in this fourth industrial revolution.

19.2 SIGNIFICANCE OF THIS CHAPTER

The significance of this chapter is that it provides managers with a sustainable framework to adopt relevant technologies as we bring them closer to digital tools in this fifth industrial revolution. Managers rely on information to drive their organizations to greater heights; however, they have been operating without a clear framework to guide them on how to adopt new emerging technologies. The chapter gives a brief background of the industrial revolutions and then introduces big data as the most strategic resource of the 21st century. The chapter makes a contribution by providing a framework that can be used by managers as a steppingstone to move from the fourth industrial revolution into the fifth industrial revolution where they are expected to be interacting with digital tools in their operations.

19.3 KEY DRIVERS OF INDUSTRY 4.0

Many researchers agree that Industry 4.0 is primarily driven by technology, including cyber-physical systems (CPS), the internet of things (IoT), embedded systems, sensors, cloud computing/services, agent-based systems, robotics, additive manufacturing, machine power, analytics, connectivity, electronic data exchange, remote operations, human-computer interaction, 3D printing, intelligent production, big data, and artificial intelligence (AI). We conducted a search for these key drivers of Industry 4.0, focusing on emerging items in each journal until we reached saturation. Table 19.1 displays the findings.

It is quite clear from the above analysis that technology will never stop emerging. Hence, there is no casting stone to pinpoint one particular one as the key driver. We then concluded that Industry 4.0 is driven by an abundance of new open-source tools/technologies that continue to emerge from IoT, suitable for knowledgeable programmers and data scientists according to their business requirements (Gokalp, Kayabay, Akyol, Eren, & Kocyigit, 2016).

19.4 DAWN OF THE FIFTH INDUSTRIAL REVOLUTION

Technology has been advancing at a faster pace than managers can cope with, to the extent that its adoption by managers has lagged behind. The fifth industrial revolution is now around the corner and is poised to correct this imbalance and bring humans closer to digital tools (Noble, Mende, Grewal, & Parasuraman, 2022). The COVID-19 pandemic sent panic waves to most managers across different sectors as they were forced to stay at home and interact with these digital tools to monitor and manage their workplaces. Those who had invested in analytics tools such as BI reaped the fruits of their investments as they continued to receive auto-generated reports from the BI systems because managers rely on information to make quality decisions. Those who choose to ignore the current demands are likely to sink.

19.5 BIG DATA ANALYTICS IN AN INDUSTRY 4.0 ENVIRONMENT

The term "big data" first appeared in 1998 (Fan & Bifet, 2015). Initially, it was seen as a challenge, with only three characteristics (volume, velocity, and variety) dominating the discussion (Kaisler, Armour, Espinosa, & Money, 2013a). Later, two more characteristics, veracity and value, were introduced. Veracity refers to a dataset that is full of uncertainty (Shukla, Yadav, Kumar, & Muhuri, 2020). Value is described as an important feature that has driven the hype for big data, as it brings added value from

TABLE 19.1 Key Drivers of Industry 4.0

Document Name	Key Drivers of Industry 4.0	References
Towards Industry 4.0: An Overview of European Strategic Roadmaps	• **Organizational:** New ways of working and collaboration. • **Technology:** Cyber-physical systems, Internet of things, embedded systems, sensors, cloud computing/services, agent-based systems, robotics, and additive manufacturing. • **Innovation:** New business models and extended innovation. • **Operational:** Performance improvements.	Santos et al. (2017)
Big Data Challenges and Opportunities in the Hype of Industry 4.0	• **Technology:** Power of machines, computing, analytics, connectivity, cyber-physical systems, internet of things, automation, cloud system, data exchange, remote operations, Human Computer interaction, 3D printing, intelligent production, big data.	Khan et al. (2017)
Big data for cyber physical systems in Industry 4.0: a survey	• **Technology:** Cyber-physical systems, big data, Internet of things, cloud computing, data science, industrial information integration, human computer interaction.	Da & Duan (2019)
Industrial Artificial Intelligence in Industry 4.0 – Systematic Review, Challenges, and Outlook	• **Technology:** Cyber-physical systems, Internet of things, ICT technologies, big data, and AI as one of the key technologies.	Peres, Jia, Lee, Sun, Colombo, & Barata (2020)
Big Data for Industry 4.0 a Conceptual Framework	Abundance of new open-source tools: that continue to emerge from internet of things suitable for knowledgeable programmers and data scientists.	Gokalp, Kayabay, Akyol, Eren, & Kocyigit (2016)

the collected data to improve processes, activities, and predictive analysis (Demchenko, Grosso, & Laat, 2013).

Nowadays, big data is known for its five characteristics: veracity, value, volume, velocity, and variety (Bello-Orgaz, Jung, & Camacho, 2016). It is defined as "a set of techniques and technologies that require new forms of integration to uncover large hidden values from large datasets that are diverse, complex, and of a massive scale" (Hashem, Yaqoob, Anuar, Mokhtar, Gani, & Khan, 2015, p. 100). In an Industry 4.0 environment, these interconnected systems and gadgets produce a large complex volume of variable data.

The hard-core sciences are interested in determining how the data will be stored and the processing performance (Heureux & Member, 2017). Other fields, such as commerce, take a keen interest in value, variety, and veracity, as they aim to extract value from a variety of trustworthy sources. To be effective in the current environment, managers must analyze this data and make quality decisions. The gadgets must be configured to report to them directly; however, they are approaching this challenge without a clear strategy for doing so, and this has been the main point of failure. "Acquiring data without analytics software to extract value is equivalent to simply improving the organization's infrastructure" (Watson, 2014, p. 1248). Hence, managers have been improving their infrastructure without deriving value from the Industry 4.0 environment.

19.6 AIM AND OBJECTIVES

The ultimate aim of this research was to develop a framework that managers can utilize to implement a data analytics strategy.

19.6.1 OBJECTIVES

- To assess managers' awareness of the benefits of interconnected devices in light of the ongoing emergence of new technologies;
- To determine whether the increase in information from disparate sources now affects operations and managers' expectations in this new environment;
- To explore the role of the IT department and external consultants in the new environment; and

- To provide a solution of what managers need to do to embrace the new demands of Industry 4.0.

19.7 METHODOLOGY AND DATA

The research adopted relativism as an ontology and then took an emic approach by immersing itself in the world of managers.

We conducted an exploratory qualitative case study that involved in-depth interviews with executives, senior managers, internal consultants, and external consultants as targeted participants. A focus group was then used to validate the findings. The idea was to explore their practical experiences with emerging technologies and interconnected gadgets.

The managers in the study have varying lengths of working experience in data analysis, utilizing manual means, Excel, enterprise resource planning (ERP) systems, and business intelligence (BI) systems. They play a crucial role in determining the strategic direction of the organization. Additionally, they are the internal "owners" of business processes involved in decisions to invest in new emerging technologies. A total of 23 participants took part in this research study, including 5 executives, 5 senior managers, 3 internal consultants, 5 external consultants, and 5 internal consultants/managers who participated via a focus group.

19.8 THEORETICAL ASSUMPTIONS

The research study was underpinned by two theories: (i) The Unified Theory of Acceptance and Use of Technology (UTAUT); and (ii) the diffusion of innovation theory (DOI) as theoretical lenses. These two theories also guided the interview questions. The Effort Expectancy of (UTAUT) came in handy because it refers to the perceived ease of use of any new system/technology, in other words, the degree to which the prospective user expects the new technology to be free of effort. The Performance Expectancy of (UTAUT) also refers to the degree to which an individual believes that using the system will help improve their performance.

The complexity determinant of (DOI) refers to the degree to which an innovation is perceived as difficult to understand and use.

The DOI theory comprises five steps: (i) knowledge; (ii) persuasion; (iii) decision; (iv) implementation; and (v) confirmation (Sahin, 2006).

These steps are useful in examining why managers are persuaded to implement new technology but do not use it.

The two theories were compared and found to share common elements such as effort, resources, and benefits (Figure 19.1).

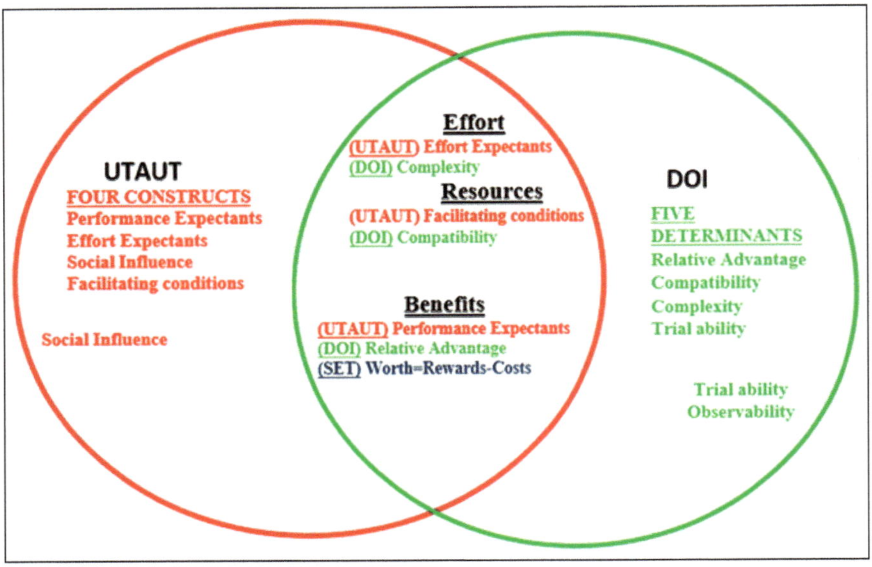

FIGURE 19.1 Engaged theories on common ground.

The UTAUT theory has four constructs, and one of these constructs, "facilitating conditions," refers to the extent to which an individual or organization believes that the organizational structure and technical infrastructure are in place to support the use of the system.

The DOI theory's complexity construct also refers to "the degree to which an innovation is perceived as difficult to understand and use" and was used to measure the complexity of integrating the relevant technologies needed to analyze information in the big data environment.

We followed Rogers' five communication stages of the DOI theory: knowledge-awareness, persuasion, decision implementation, and confirmation. The assumption was that since the managers have made a decision to adopt and implement other technologies like the ERP and BI systems, they have moved beyond the persuasion stage and have developed a positive attitude towards the innovation. However, the question remains: do they trust these new technologies and do they wish to continue investing in them?

Do managers possess knowledge and awareness of other emerging technologies that generate new sources of information and the associated advantages in terms of relative advantage? Do they believe that facilitating conditions, such as organizational structure and technical infrastructure, are in place to leverage these new sources? These questions guided the interview.

19.9 RESULTS AND DISCUSSION

The analysis in this research followed a thematic approach based on themes that emerged during data collection.

The main theme was what managers need to do to benefit from emerging technologies in this fourth industrial revolution, and the following were the sub-themes: (i) benefits of using interconnected systems and devices; (ii) managers' expectations; and (iii) the role of the information technology department and consultants.

For each sub-theme, we re-examined the theoretical assumptions, then compared them to the outcome of semi-structured interview questions that we asked during the interviews.

19.9.1 OUTCOME FROM THE BENEFITS OF USING INTERCONNECTED SYSTEMS AND DEVICES THEME

The managers are aware that we are currently living in a digital world. They understand that investing in the latest technologies can serve as strategic tools to effectively respond to competition. It is crucial for them to have analytics solutions in place in order to maintain competitiveness in the current business environment. These solutions provide valuable insights into the present, future, and past states of the business. Consequently, the managers are not overwhelmed by the emergence of new technologies. However, they face technical limitations when it comes to selecting the most suitable solution, which is why they always seek guidance from the IT department. The key question then arises: who should take the lead in digitalization during the fourth industrial revolution, the managers as process owners or the IT department as technical owners?

Managers know that interconnected systems bring benefits. For example, the vehicle tracking system adds significant value to the organization, especially when looking at the drivers' behavior. Drivers handle one of the organization's most valuable pieces of equipment and they take it out of sight. They go to places where the manager may not know and don't even see them, so the vehicle tracking system, using sensors, becomes the managers' only source of comfort to understand the driver's behavior. This is also a cost-saving measure that can be used by the manager to manage the performance of the trucks, the drivers, and fuel consumption. It can then be used as a balanced scorecard to measure the driver's performance.

In manufacturing, interconnected systems will enable buyers to track their orders and engineers to verify spare availability at the source before placing orders. By utilizing machine sensors, bar codes, and scanners, managers will have precise knowledge of production output and the ability to track products to the market. However, despite the numerous benefits associated with these technologies, their adoption has remained a topic of discussion in most organizations' boardrooms (EY, 2014). Managers lack a clear strategy for implementing a big data analytics strategy, resulting in a slow uptake.

19.9.2 OUTCOME FROM THE MANAGERS' EXPECTATIONS THEME

Data sources play a significant role when an organization embarks on an analytics journey. Managers desire an e-infrastructure with drill-down capabilities to access and analyze all the information in the environment, enabling them to identify the root cause of problems and make informed decisions. In the manufacturing industry, automated breakdown reports are sought after to aid in determining when to service machines before they experience failure.

They also want to automate manual processes to reduce the number of people involved in non-value-adding tasks. The COVID era has prompted a change in their mindset, and they desire automation to be implemented as soon as possible. Their request to the IT team is to internalize the automation process, utilizing internal staff who have a better understanding of the business processes. They suggest having knowledgeable managers within each department who can serve as key liaisons and provide suggestions for system improvements. This approach aims to avoid a situation where consultants create elaborate solutions that fail to address the department's actual problems. To ensure success, all managers within the department must collectively tackle the challenges. They firmly believe that with proper configuration, automation will enable seamless information flow and deliver highly credible results.

However, as they scale up the technology of interconnecting these systems, managers have no confidence in the security systems. They expect those interconnecting the systems to put security measures in place to safeguard against cyber-attacks, leakage of strategic information, and formulas. They are afraid that someone can hack the systems and be able to place fictitious orders and withdraw stocks without paying. To overcome this fear, there is a need to prevent unauthorized access.

19.9.3 OUTCOME FROM THE ROLE OF THE INFORMATION TECHNOLOGY DEPARTMENT AND CONSULTANTS THEME

The manager is the focal point as the owner of the process; however, the roles of both the external consultants and the IT department are still relevant. The modeling and integration require someone from IT or someone who is really knowledgeable about the terminologies of the emerging technologies. The IT department also plays a key role as a stakeholder in terms of specifications and assists in choosing the appropriate technology. They are the ones who, upon consulting with the business, would determine the best tool for you based on your requirements, volumes of data, and strategic plan. They are responsible for sourcing and administration of hardware such as servers, switches, sensors, and cameras. They then set up the environment or the network infrastructure and grant the necessary authorizations. If there are skilled in-house personnel, they lead user departments on how to carry out the implementation. The external consultants cannot be eliminated because they are the owners of the solutions. They know how these solutions operate.

19.10 PROPOSED FRAMEWORK FOR ADOPTING A DATA ANALYTICS STRATEGY

A data analytics strategy requires collaboration from all departments along the value chain. The entire process of crafting and adopting a data analytics strategy should involve managers from different departments, not just the traditional IT department. With the emergence of new technologies in this fourth industrial revolution, organizations can no longer rely solely on one technology, such as ERP systems, to cover the entire value chain for transactional purposes. There is now a clear need for managers to be innovative and automate all processes along the value chain using the available technologies.

We have developed the concept of the "bucket of emerging technology" which enables managers to scan the bucket and select the appropriate technologies they need for their operations. There are five steps involved:

- The manager must first examine the current processes to identify processes that can be automated.
- Scan the bucket of emerging technologies to choose the appropriate technologies that suit their operations. If the manager cannot find the technology, they must be innovative to develop one by seeking help from the developers.
- A decision must be taken to consume the technology on premise or from the cloud.
- Source for an implementation partner plan to integrate with existing technologies and budget to implement.
- Implement the new technology and integrate it with the existing one.
- Please scan the bucket of emerging technologies again to select the best available big data analytics system or software, which may also utilize a data warehouse or analyze data directly from the source.
- Implement the big data analytics software.

The above processes 1 to 7 are repeated regularly to implement new emerging technologies or replace the existing ones because new technologies will never stop emerging. A diagrammatical representation of the bucket of emerging technologies concept is given in Figure 19.2.

Technology will never stop emerging, and those who do not embrace it will be left on the sidelines. There is a need for managers to always scan the environment for new emerging technologies. They can seek the help of the IT department for further clarification, but the focal point is the manager because the manager knows the pain points in operations. They are also knowledgeable about the internal business process to guide innovation. Following the value chain concept, managers should be given the chance to look for the best solutions and pick their preferences from those available.

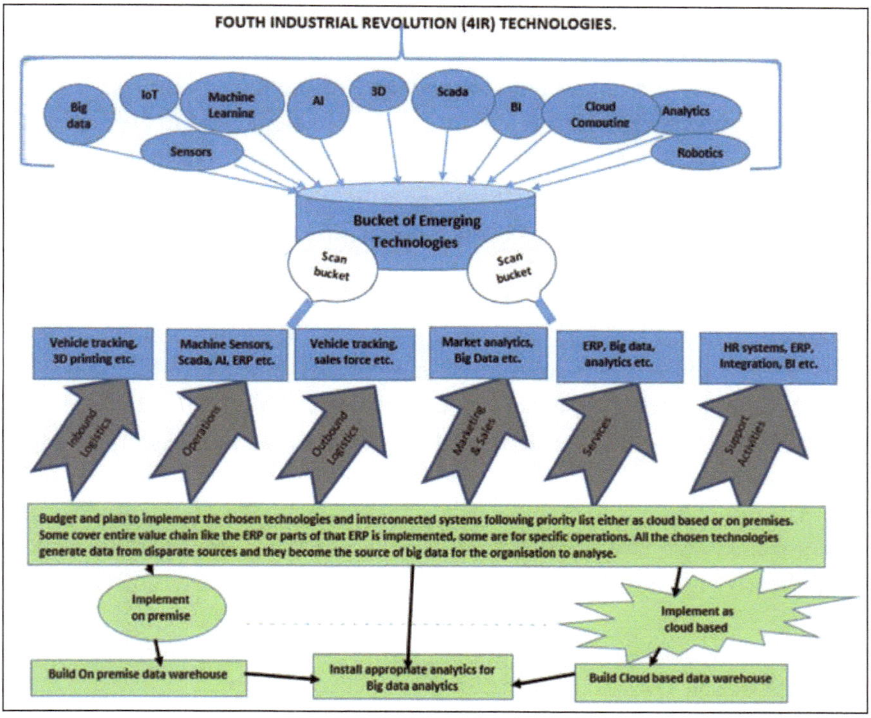

FIGURE 19.2 Data analytics' strategy.

Literature identifies big data as a byproduct of business transactional systems such as ERP (McDonald & L'eveill'e, 2014). Data from machine sensors, vehicles stationary and in motion (Matti & Kvernvik, 2012). Data produced by public service institutions (Administrative data). Data from the IoT, social media, and networks such as Google, Facebook, Twitter, LinkedIn, etc. (McDonald & L'eveill'e, 2014). Data from sources that breed unstructured, semi-structured, and structured data (Baesens, Bapna, Marsden, Vanthienen, & Zhao, 2016).

19.11 CONCLUSION

We can summarize it by stating that big data is a byproduct of emerging technologies. The fourth industrial revolution is driven by Cyber-Physical Systems (CPS) utilizing various emerging technologies, which continue to emerge. These interconnected systems generate vast amounts of information at a rapid pace, making manual analysis impractical. As a result, there is a growing need to configure systems, machines, gadgets, etc., to directly report to managers, forming the foundation of big data analytics.

To ensure success, sufficient time should be dedicated to evaluating the pool of emerging technologies in order to select the most suitable solutions for operations and identify the best implementation partners. In many cases, projects fail due to either inadequate solutions or unsatisfactory implementation partners.

Managers should then budget to implement the chosen technologies in their operational budgets. Those technologies that cover the entire value chain, such as ERP systems and BI analytics systems, should be budgeted for under the organization's consolidated master budget. A decision should then be made to either implement the solutions on-premises, in the cloud, or both.

The final phase is to scan the pool of emerging technologies again to look for a robust data analytics system, such as a BI system or audit analytics, and identify the best implementing partner for the analytics software.

KEYWORDS

- **analytics software**
- **big data**
- **business intelligence**
- **cyber-physical systems**
- **data analytics strategy**
- **fifth industrial revolution**
- **Industry 4.0**

REFERENCES

Al-Habaibeh, A., Watkins, M., Waried, K., &Javareshk, M. B., (2021). Challenges and opportunities of remotely working from home during COVID-19 pandemic. *Global Transitions* (Vol. 3, pp. 99–108). The Authors.doi: 10.1016/j.glt.2021.11.001.

Alharthi, A., Krotov, V., & Bowman, M., (2017). Addressing barriers to big data. *Business Horizons* (Vol. 60, No. 3, pp. 285–292). Kelley School of Business, Indiana University. doi: 10.1016/j.bushor.2017.01.002.

Baesens, B., Bapna, R., Marsden, J. R., Vanthienen, J., & Zhao, J. L., (2016). Transformational issues of big data and analytics in networked business. *MIS Quarterly, 40*(4), 807–818. doi: 10.5121/ijgca.2012.3203.

Bello-Orgaz, G., Jung, J. J., & Camacho, D., (2016). Social big data: Recent achievements and new challenges. *Information Fusion* (Vol. 28, pp. 45–59). Elsevier B.V. doi: 10.1016/j.inffus.2015.08.005.

Da, L., & Duan, L., (2019). *Big Data for Cyber Physical Systems in Industry 4. 0: A Survey, 13*(2), 148–169.

Demchenko, Y., Grosso, P., & Laat, C. D., (2013). Addressing big data issues in scientific data infrastructure. *Proceedings of the 2013 International Conference on Collaboration Technologies and Systems, CTS 2013* (pp. 48–55). doi: 10.1109/CTS.2013.6567203.

Fan, W., & Albert, B., (2014). Mining big data: Current status, and forecast to the future for telecom data. *International Journal of Private Cloud Computing Environment and Management, 1*(1), 1–10. doi: 10.21742/ijpccem.2014.1.1.01.

Gokalp, M. O., Kayabay, K., Akyol, M. A., Eren, P. E., & Kocyigit, A., (2016). Big data for industry 4.0: A conceptual framework. In: *International Conference on Computational Science and Computational Intelligence* (pp. 431–434). doi: 10.1109/CSCI.2016.87.

Grossman, R. L., & Siegel, K. P., (2014). Organizational models for big data and analytics. *Journal of Organization Design, 3*(1), 20–25. doi: 10.7146/jod.9799.

Hashem, I. A. T., Yaqoob, I., Anuar, N. B., Mokhtar, S., Gani, A., & Khan, S. U., (2015). The rise of "big data" on cloud computing: Review and open research issues. *Information Systems, 47*, 98–115. Elsevier. doi: 10.1016/j.is.2014.07.006.

Heureux, A. L., & Member, G. S., (2017). Machine learning with big data: Challenges and approaches. *IEEE Access, 5*, 7776–7797. IEEE doi: 10.1109/ACCESS.2017.2696365.

Kaisler, S., Armour, F., Espinosa, A., & Money, W., (2013a). Big data: Issues and challenges moving forward. In: *Proceedings of the Annual Hawaii International Conference on System Sciences*, (pp. 995–1004). doi: 10.1109/HICSS.2013.645.

Khan, M., Wu, X., Xu, X., & Dou, W., (2017). Big data challenges and opportunities in the hype of Industry 4.0. *IEEE International Conference on Communications*. doi: 10.1109/ICC.2017.7996801.

Korhonen, J. J., (2014). Big data – big deal for organization design? *Journal of Organization Design, 3*(1), 31–36. doi: 10.7146/jod.13261.

Noble, S. M., Mende, M., Grewal, D., & Parasuraman, A., (2022). The fifth industrial revolution: How harmonious human–machine collaboration is triggering a retail and service [R] evolution. *Journal of Retailing, 98*(2), 199–208. Elsevier Inc. doi: 10.1016/j.jretai.2022.04.003.

O'Reilly, C. A., & Roberts, K. H., (1974). Information filtration in organizations: Three experiments. *Organizational Behavior and Human Performance, 265*(1), 253–265.

Peres, R. S., Jia, X., Lee, J., Sun, K., Colombo, A. W., & Barata, J., (2020). Industrial artificial intelligence in Industry 4. 0–systematic review, challenges, and outlook. *IEEE Access, 18*, pp. 220121–220139.

Sahin, I., (2006). Detailed review of rogers diffusion of innovations theory and educational technology-related studies based on rogers. *The Turkish Online Journal of Educational Technology, 5*(2), 14–23.

Santos, C., Mehrsai, A., Barros, A. C., Araujo, M., & Ares, E., (2017). Towards Industry 4.0: An overview of European strategic roadmaps. *Procedia Manufacturing, 13*, 972–979. Elsevier. B.V. doi: 10.1016/j.promfg.2017.09.093.

Shukla, A. K., Yadav, M., Kumar, S., & Muhuri, P. K., (2020). Engineering applications of artificial intelligence veracity handling and instance reduction in big data using interval type-2 fuzzy sets. *Engineering Applications of Artificial Intelligence* (Vol. 88, p. 103315). Elsevier Ltd. doi: 10.1016/j.engappai.2019.103315.s

Snyder, S. A., & Barnakova, Y., (2020). Being a digital leader has never been more urgent. *Knowledge at Wharton: A Business Journal from the Wharton School of the University of Pennsylvania*. Available at: https://knowledge.wharton.upenn.edu/article/theres-never-important-time-digital-leader (accessed on 11 January 2023).

Index

A

Academic implication, 102
Additive manufacturing, 89, 92, 239, 240, 303, 305
Administrative data, 313
Agricultural fields, 171
Agroeconomy, 273
Agronomy, 96
Agrotourism, 179
Air connectivity, 177, 179, 182
Airtel, 180
Algorithm training, 211
Alternative hypothesis, 44, 45
Amazon, 39, 46, 113, 118, 121
 web services (AWS), 113, 121
Ambiguity, 88, 101
Analytic hierarchy process (AHP), 111
Analytical techniques, 38
Analytics software, 306, 312, 314
Animal husbandry, 179
ANOVA, 40, 44, 46, 161, 166
Antarctica, 65, 69
Artificial intelligence (AI), 22, 27, 28, 64, 88, 89, 91, 99, 237, 304, 305
Artisans, 34, 36–40, 45–47, 49–52, 178, 181
Audio systems, 206
Audit standards, 155, 156
Augmented
 Dicky Fuller (ADF), 131
 reality, 88, 89, 92, 206, 208, 240
Australian public sector, 144
Auto regressive distributed lag (ARDL), 125, 131, 132, 135
Autocorrelation, 134, 136
Automated
 production lines, 91
 teller machine (ATM), 125, 126, 128, 130–133, 135, 136
Automation, 88, 90, 92, 99, 102, 110, 127, 129, 130, 136, 137, 305, 310, 311
Automobiles, 185, 186, 190, 191, 240
Autonomous robot, 89, 92, 239, 241
Auto-regressive integrated moving average (ARIMA), 110
Average variance extracted (AVE), 280, 281
Awareness, 47, 50, 52, 60, 68, 69, 103, 163, 202, 206, 221, 222, 227, 229, 230, 244, 265, 289, 306, 308, 309
Aymeric Damien's neural network example, 11

B

Bamboo
 crafts, 33, 34
 processing, 177
Bandwidth, 117
Banknotes, 7
Bibliographic database, 204
Bibliometric analysis, 204
Big
 data, 27, 58, 62–64, 67–69, 89, 91, 92, 208, 210, 240, 302–306, 308, 310, 312–314
  analysis, 58, 69
  solutions, 67, 68
  technologies, 62
 three, 118
Bioeconomy, 27, 90, 91, 96, 102
Bioenergy, 96
Bio-fertilizer unit, 181
Biogas plant, 181
Biological resources, 90, 96
Bio-products, 96
Biotechnology, 96
Bitcoin, 1, 4–14
 blockchain, 5

characteristics, 6
bitcoin user's profile, 8
political neutrality, 8
shared market, 7
community, 5, 6, 8, 9
future, 9, 10
history, 5
magazine, 5
market, 4
prices, 5
reserves, 4
traders, 4
transactions, 7, 10, 12, 14
users, 5, 8, 9
Black pottery, 38
Blatant misinformation, 64, 69
Blockchain, 6, 10, 11, 23, 241
Bodoland Territorial Council (BTC), 10, 178
Bretton woods system, 4
Breusch Pagan Godfrey Test, 134
Bricolage, 218, 222, 224, 225, 227, 229
British Pound, 4, 6
Bronze coins, 7
Business
canvas, 211
intelligence (BI), 114, 304, 307, 308, 314
positioning, 201
process monitoring, 25
resilience, 24
transactional systems, 313
transactions, 127

C

Capital market regulator, 141
Cashless financial transactions, 126
Cellular technology, 130
Census data, 75
Chaos, 59, 61, 100, 101
Chief Executive Officer (CEO), 146
Chikankari craft, 37, 38
Chinese RMB, 4
Chi-square test, 78, 83, 243
Circular intelligent products, 24
Civil
infrastructure, 182
society, 218
Clearinghouse, 34, 35
Cloud
analysis, 69
based applications & resources, 111, 114, 116
characteristics, 111
high availability, 112
large-scale networking, 112
pay as you go, 112
resources pooling, 111
secure, 112
self-service on the go, 112
simple maintaining, 112
classifications, 112
cloud services types, 114
computing, 27, 89–91, 107, 108, 110–117, 119, 120, 240, 303, 305
stack, 114
suppliers, 116
data gathering, 127
experts, 111
foundry environment, 108
management, 110, 117, 118
platform, 110
services, 110
manufacturing, 23, 24
networks, 108
platform, 108, 111, 121
portfolio management operations, 117
professionals, 110
provider, 111, 117, 118
servers, 90, 109
service, 114, 116–120
companies, 116
provider, 107, 120, 121
storage providers, 112
technology, 22
transparency, 117, 118, 121
CoinDesk, 10
Collaborative robots (CoBot), 23, 89
Communication, 22, 61, 89, 92, 96, 165, 179, 201, 203, 205, 206, 225, 239, 240, 245, 261, 264, 282, 308
Compare cloud service providers (CSPS), 107–111, 113–118, 120
industry challenges, 116
transparency advantages, 117

Complex dynamic systems, 109
Complexity, 88, 100
Composite reliability (CR), 280, 281
Computing resources, 108, 111, 113
Consensus
method, 8
system, 6
Consumer
rights investigation forum, 77
satisfaction, 77
Consumption, 9, 26, 73, 77, 85, 192, 193, 256, 310
Continent, 35, 69, 242
Contract conditions, 118
Conventional
human-computer interface, 205
reporting, 143
thinking, 165, 171, 218
Corona viruses, 46, 127
Corporate social responsibility (CSR), 143
Cost-effective provider, 117
Cottage industry, 47
Counter infodemic strategies, 67
COVID, 21, 33, 34, 36, 39, 41, 42, 52, 57, 59–64, 68, 69, 73–78, 80, 83–85, 127, 140, 155, 225, 230, 246, 271–278, 280–283, 301, 302, 304, 310,
misinformation, 64, 69
vaccination, 74
Craft, 37, 38, 42, 47, 48, 51
Crisis management, 68
Critical success factors, 217, 218, 224, 225
Cryptocurrencies, 1, 3–6, 10, 11, 14
Cryptographic proof, 5
Cryptography, 4
Cumulative sum (CUSUM), 134
Currencies, 1, 3, 4, 6, 7
Customer life cycle, 90
Customization, 90
Cutting-edge robotic technology, 30
Cybercrimes, 99
Cyber-physical
cognitive systems, 88
systems (CPS), 88, 89, 91, 245, 301–303, 305, 313, 314
Cybersecurity, 89, 92, 103, 240

D

Data
analysis, 28, 40, 79, 85, 135, 137, 240, 245, 246, 262, 263, 269, 278, 283, 299, 307
techniques, 263
tools, 79, 85
analytics, 62, 68, 89, 240, 302, 306, 310–312, 314
strategy, 306, 310, 311, 314
availability, 196
breaches, 99
centers, 51, 109, 118
collection techniques, 262, 269
dashboards, 64
description, 130, 137
intelligence, 63
interoperability, 28
interpretation, 135, 137
sharing, 25
sources, 10, 62, 85, 130, 165, 262, 263, 269
statistics, 11
Database, 28, 97, 108, 114, 126, 130, 204
Dataset, 10–13, 64, 69, 134–136, 304
Decision-making, 99, 110, 144, 145, 149, 153, 155, 156, 218, 258, 261, 292, 293
Deflation, 9, 10
Demographic profile, 79, 147, 279
Demonetization, 39, 129
Design
cell, 191, 192
India, 47
innovators, 195
management, 186, 188–190, 195, 196
Development policy, 255, 257–259, 264, 266, 268
Digital
age, 48
Assam, 180
bank accounts, 137
business models, 89
club, 180
economy, 125, 127–130, 137
financial services, 129
gig workers, 53

India, 179, 180
information systems, 59
infrastructure, 175, 179, 182, 183
literacy, 137, 180
marketing, 46, 50
media, 164
northeast and digital Assam, 183
patent data, 95
payments, 125, 126
platform, 180
publicity, 50
skills, 34
technology, 126, 182, 201, 242, 245
transactions, 126–128
transformation, 127, 239
twins, 23
Digitalization, 22, 126, 128–130, 136, 137, 238, 241, 242, 302, 309
Digitization, 100, 102, 125, 237, 238
Dima Hasao
Autonomous Territorial Council (DHATC), 178
Territorial Council (DHTC), 178
Direct electronic exchanges, 5
Disaster management, 63
Disease transmission, 76
Double spending, 6
Drill-down capabilities, 310
Dynamic
conditional correlations, 6
modeling, 28

E

Eco-friendly
environment, 183
tourism development, 175
Econometric
analysis, 131
modeling, 125
models, 130, 135
Economic
balance, 8
costs, 274
development, 3, 7, 23, 175, 179, 181, 182, 218, 220, 226–229, 231, 257
growth, 23, 24, 34, 126, 129, 162, 227, 256, 257, 259, 264
inequalities, 224
issues, 224, 227
power balance, 3
utility, 221
Ecotourism, 179
Edge computing, 23
Education sector, 179
Educational institutions, 271–276, 282
E-governance, 182
programs, 183
E-infrastructure, 310
Electrocardiogram (ECG), 206
Electroencephalogram (EEG), 206
Electronic
cash system, 5
devices, 240
domains, 191
financial transactions, 126
sectors, 185, 186
supplies, 48
transactions, 5
E-marketing, 47
Emergent intelligence, 99
Emotional intelligence, 241
End-to-end networking, 238
End-user, 119
corporations, 117
organizations, 117
Energy
commodities index, 6
industries, 96
Enterprise resource planning (ERP), 22, 307, 308, 312–314
Entrepreneurial
ecosystem, 227
passion, 223
spirit, 227
Entrepreneurs, 33–37, 40, 46, 47, 49–52, 181, 183, 217, 218, 220–230
Entrepreneurship, 34–36, 221, 222, 224, 227, 230
Environmental
conditions, 211
environmental, social, and governance (ESG), 141, 142

factors, 77
issues, 23, 188
Error correction model, 135
Essential oil extraction, 181
Ethnic group identity, 222
Euro, 2–4, 6
European
Commission, 23, 96
Union (EU), 2, 4, 8
Evolution, 87, 88, 97, 103, 140, 211, 239, 242
Exponential growth, 7, 60, 69

F

Fact-checking mechanisms, 63
Female
enterprises, 36
entrepreneurs, 34, 36, 48
entrepreneurship, 34
Fifth industrial revolution (Industry 5.0), 1, 21–30, 33, 57, 73, 87–92, 95, 97–103, 107, 125, 127–130, 137, 139, 156, 161, 175, 185, 200, 201, 217, 237, 238, 242, 244–246, 252, 255, 271, 284, 289, 301–304, 314
key challenges, 98
Finance industry, 38
Financial
assistance, 38, 177
ecosystem, 218
inclusion, 125–130, 136, 137
institutions, 5, 9, 170
market, 3, 7, 8, 14, 125–129, 137
resources, 224, 225, 230
sector, 125, 127, 129, 136
transactions, 125, 127, 129, 130, 137
Firm
gross turnover, 191
machinery, 191
Fishing, 96
Flipkart, 39, 46
Food
Food and Agriculture Organization (FAO), 77
manufacturing, 29
sciences, 96
testing facility, 181
Foreign direct investment (FDI), 162, 189
Forestry, 96
Fourth industrial revolution (Industry 4.0), 23, 26, 27, 30, 88–92, 95, 97, 103, 129, 237–240, 242–246, 289, 299, 301–307, 314

G

Gender
disparity, 45
equality, 35
General commodity index, 6
Genesis block, 5
Genetic algorithm, 109
Geographical
analysis, 65, 69
location, 2, 182
map, 65, 66, 69
Geo-location mapping (GM), 206
Gig
terminology, 46
workers, 39, 46, 48
Global
crisis, 44, 221
currency, 1–4, 7–9, 14
digital platforms, 241
economic value, 240
handicraft
index, 39
market, 39
market, 5, 9, 187
navigation satellite system (GNSS), 206
organizations, 2
political discussions, 76
reporting initiative, 141
Globalization, 1, 14, 38, 140, 155, 241
Gold, 3–7, 9, 14, 166–168, 171
standard convention, 5
Good reporting practices, 144
Google, 97, 118, 202, 203, 275, 313
Government initiatives, 23
Granger-causality, 131, 133, 136
test, 133
Graphical lasso regularization network, 109
Green
energy, 183
vegetables, 73, 85

H

Handcraft
businesses, 38
management, 39
workers, 45
Handicraft, 34, 38, 39, 50–52, 177, 178
business, 38, 45
education, 52
enterprises, 39
industry, 36, 38, 40, 46, 48, 49, 51, 52
market, 48, 129
parks, 48
products, 36, 37, 39, 48, 50
sector, 33, 36, 37, 39, 40, 43, 48, 49, 52, 53
workers, 41, 42, 44
Handloom, 38, 47, 48, 178
Handmade
businesses, 35
carpet industry, 39
industry, 36, 40
paper unit, 181
pottery, 46
products, 38, 39, 47, 48
sector, 35, 39, 47, 50
Health
effects, 85
service centers, 84
technicians, 83, 84
Healthcare, 63, 74, 88, 225, 227, 237, 243, 246
status scale, 243
systems, 63
Healthy
bubble, 9
organizational climate, 67
Hedge, 6
Higher market capitalization, 10
Hindustani craftworks, 38
Horticulture, 177
Human
centricity, 25, 26
factor, 26
population, 57
resources, 102
robot co-working, 97, 100, 102, 103
collaboration, 91, 96
-to-machine communication, 89
Humanization, 265
Hybrid cloud, 113
services, 113
Hydraulic energy, 183
Hydroxychloroquine, 60
Hypothesis testing, 45, 278

I

Implementation strategy, 50
Indian
craft business, 38
economy, 52, 125, 128, 131, 136, 137, 189
Institute of Management & Technology, 51
market, 127
Industrial
innovation, 185, 186, 194, 245
internet of things (IIoT), 89, 92, 99, 239, 240
landscape, 189
production, 26
revolution, 23, 87, 88, 91, 97, 103, 238, 239, 241–243, 301–303, 309, 312, 313
sector, 87, 238, 239, 243
technologies, 92
Industrialization, 239
Industry
1.0, 87, 88, 91, 302
2.0, 88, 91
academic cooperation, 190
services, 242
Infodemics, 63, 67–69
situation, 59
Information
and communication technology (ICT), 22, 92, 96, 175, 177, 180, 182, 183, 211, 305
tools, 22
fatigue, 63
glut, 57, 58
management, 187, 201
technology, 59, 87, 96, 102, 107, 175, 180, 244, 302, 309
transmission, 59

Infrastructure as a service (IaaS), 114, 117, 121
Initial internal market, 2, 7
Innovation
chain, 187
strategy, 187
Innovative
digital technologies, 91
India, 47
industrial designs, 245
Institutional framework, 52, 53
Integrated
business model, 209, 213
reporting (IR), 139–146, 149–156
adoption, 144, 156
Integration, 1, 22–25, 28, 89, 91, 92, 131, 135, 154, 193, 201, 239, 244, 245, 305, 306, 311
Intellectual capital (IC), 97, 143
Intelligent
automation, 24, 25, 89
healthcare, 23, 24
rules, 109
wearables, 213
Intermediary currency, 4
International
capitalism, 258
conferences, 49
Integrated Reporting Council (IIRC), 140–142, 154
trade, 2, 3
Internet
banking, 125
connectivity, 126, 180, 182, 240, 276
of everything, 23
of services (IoS), 89
of things (IoT), 27, 89, 91, 99, 126, 202, 206, 208, 210, 237, 238, 241, 245, 303–305, 313
Inter-organizational innovations, 92
Interpersonal relationships, 225, 226, 229, 245
Intra-organizational collaboration, 92
Investment
guidelines, 165
practices, 162, 164–170
Investor relations, 143–145, 154–156

J

Japanese Yen, 4, 6
Jardozi, 38
Journal, 40, 97, 98, 204, 205, 220, 261, 291
indexing, 97
search strategy, 97

K

Kalimati Fruit and Vegetable
Market Development Board (KFVMDB), 74
Wholesale Market Development Board (KFVWMDB), 75
Karbi Anglong Autonomous Territorial Council (KAATC), 178
Kathmandu market, 85
Know your customer (KYC), 127
Knowledge management, 28
Kubernetes, 115

L

Lagrange-multiplier (LM), 134
Lamination, 181
Large-scale companies, 139, 146
Leadership, 98, 101, 142, 230, 243
Learning and professional sports domain, 206
Legal
interventions, 67, 68
measures, 68, 69
strategies, 68
Liberalization, 189
Life sciences, 96
Limited market intelligence, 52
Liquidity, 2–4, 6
Livelihoods, 36, 47, 76, 77, 266, 267
Local
community empowerment, 263, 269
employment opportunities, 36
Logarithmic transformation, 131
Long-run
benefits, 167, 168
equality, 136
equilibrium, 132
stability, 132

Long-term
benefits, 166, 167, 170
discounts, 120
investment, 171
plans, 170
viability, 118
Look east policy, 182

M

Machine
cognition, 27, 28
learning, 10–12, 68, 208, 237, 245
made products, 39, 46
sensors, 310, 313
Macroeconomic factors, 136
Malfunction, 99
Malnutrition, 74
Managerial implication, 102, 103
Manufacturing industry, 238, 310
Marginalization, 256
Market
driven strategy, 213
stability mechanism, 110
volatility, 7
Mass
customization, 88
media, 59
production, 88, 90, 91, 302
Maternal mortality, 78
Mechanical loom production, 91
Medical/healthcare features (HM), 206
Medium
enterprises, 36
scale companies, 146
Metonymic conspiracy theories, 61
Metropolitan areas, 84
Microbusinesses, 33
Microchip implantation, 213
Micro-enterprises, 34, 36, 53
Micro-entrepreneurial operations, 34
Micro-handmade, 35
Microsoft Azure, 113
Miners, 4, 6, 8, 9
Misinformation, 58–65, 67–69
Misleading, 57–61, 65, 67, 68
information, 57–59, 61, 67, 68
Mobile
financing, 126
subscriptions, 125, 126, 128, 131–133, 135, 136
Modern business practices, 172
Money scarcity, 168
Moonj crafts, 33, 34
Motivational factors, 221, 222
Multi-agent systems, 99
Multi-cloud
environments, 109
system, 109
Multi-criteria decision-making (MCDM), 109
Multinational corporations, 120
Multiple regression models, 166

N

Nanotechnology, 96
Nash equilibrium, 110
National
banks, 164
capital region (NCR), 185, 186, 190, 196
economies, 34
electronic fund transfer (NEFT), 125
index, 39
shutdown, 37
standards, 177
Nationalism, 256
Natural resources, 26, 161, 182, 265
NCapture extension, 59
Near
field communication (NFC), 213
home delivery sector, 48
Neonatal deaths, 78
Neo-structuralist
development paradigm, 258
paradigm, 258
Network
coverage, 126, 275
management, 190
Neural network, 10, 11, 13, 14
Nicotine, 202
Nike, 39, 203
Non-agricultural labor force, 35

Non-cooperative game paradigm, 110
Non-energy commodities index, 6
Non-financial
disclosures, 140
reporting, 144, 156
reports, 149, 150, 153
results, 144
Non-participatory development, 259
Non-profit
enterprises, 229
organizations, 230
venture, 229
Non-reversible transaction, 5
Northeast
road sector development scheme (NERSDS), 178
special infrastructure development scheme (NESIDS), 177
Northeastern Regional Agricultural Marketing Corporation (NERAMAC), 178
Null hypothesis, 44, 46, 296
Nutritional values, 77
NVivo, 59, 64–66, 69

O

One district one product (ODOP), 33, 34, 36, 47, 52, 53
Online
application, 109
payment, 4, 6, 7, 47
market segment, 4
system, 6
peer-to-peer transactions, 5
resources, 64
teaching, 274, 275
Ontology, 99
Orchestrators, 115
Organization building, 224
Organizational
factors, 100
forms, 241
growth, 222
practices, 226
Over-injection, 9

P

Pandemic, 33, 34, 36, 39–46, 52, 57–61, 63, 64, 68, 73, 74, 76, 77, 127, 140, 155, 225, 271–274, 276, 277, 281–283, 302, 304
Paper money, 3, 7
Partial least square (PLS), 271, 278–282
PayZapp, 127
Pediatric healthcare, 78
Peer-to-peer electronic cash system, 14
Peripheral healthcare institutions, 74
Permanent markets, 77
Personal protective equipment (PPE), 74, 84
Personalization, 25, 88, 90
Pharmaceutical, 30, 146, 185, 186
manufacturing sector, 30
Phillips Perron (PP), 131
Physical server chip technology, 119
Platform as a service (PaaS), 114, 121
Pockets, 127
Policy
implementation, 259–261, 269
makers, 22, 260
paradigm, 257, 264
Policymakers, 64, 257, 262, 276
Political
decisions, 260
landscape resolution, 222
power, 4
Postal systems, 162
Post-paid subscriptions, 130
Post-truth politics, 58
Poverty
alleviation, 227–229
reduction, 34
Practical encouragement, 212
Preferred reporting items for systematic reviews and meta-analyses (PRISMA), 204, 205, 207
Prevention, 27, 60, 69
strategies, 67
Price-performance analytics, 117
Primary data sources, 263
Private
clouds, 113
network, 113

sector entrepreneurial programs, 34
Product
design, 28, 51, 213
innovation, 28, 185, 195
Programmable logic controller (PLCs), 88, 91
Programmed electronic devices, 88
Prototypes, 195
Prototyping, 90
Public
clouds, 113, 114
consultation, 265
engagement, 69
health messages, 64
mobile data services, 130
P-value, 44, 45
Python, 11, 12

Q

Qualitative
approach, 261
factors, 130
research, 261
Quantitative
methods, 161
research, 165

R

Radio-frequency identification (RFID), 213
Randomized weighted majority algorithm, 110
Real-time
business information, 114
communication (RTC), 25
gross settlement (RTGS), 125
information, 64
warehouse inventory, 29
Recycling, 26
Refine policy definition, 144
Remote learning systems, 274
Renewable
energy, 90
integration, 24
Research
and development (R&D), 186, 187, 189, 194, 195, 203, 206, 211
data sources, 262
Resilience, 26
Resource leveraging, 224, 225, 229
Retail banking, 128, 136
Revenue generation, 22
Risk
assessment, 149, 153, 165
evaluation, 154
investments, 168, 169
tolerance, 161–163, 165, 167, 168, 172
Robotic, 26, 30, 91, 99, 303, 305
agent factors, 100
malfunction, 96
Robust digital institution, 39
Rural
community, 37, 162
environment, 164
housing, 162
investments, 162
population, 127, 171
tourism, 179
urban data, 130
women, 52, 161–172

S

Safe-haven, 3, 4, 6
Sampling technique, 40, 291
Samsung, 202, 203
SAP cloud platform, 108
Scalability, 90
Science and technology intervention north-eastern region (STINER), 181
Scientific disciplines, 96
Second
choice sources, 193
technological revolution, 238
Secondary
data, 21, 40, 78, 79, 130, 165, 263, 291
sources, 263
markets, 136
sources, 58, 165
Security, 47, 67, 90, 98, 100, 107, 109, 114, 162, 178, 183, 207, 239, 240, 243, 276, 311
Sedimentation, 267
Self-actualization, 230, 256, 266
Self-efficacy, 222, 223, 228, 229

Self-help group (SHG), 38, 40, 48, 51
Self-organization, 255
Self-realization, 256
Self-service exposure, 118
Semi-structured
  data, 62, 313
  qualitative interviews, 78
Semi-urban areas, 33
Sensational political issues, 256
Sensitization, 68
Sensor
  analyses, 27
  technologies, 28
Sericulture, 177, 179
Serverless computing, 114, 115
Service
  orientation, 25
  providers, 107, 109, 116–118, 120, 202
  sector software industries, 136
Short-run
  coefficients, 135
  directional causality, 136
  disturbances, 132
  error correction model, 131, 133
  external shock, 127
Short-term
  investments, 171
  shocks, 126
Silver, 7
Simple additive weight (SAW), 111
Skill
  development, 181
  training, 38
SkLearn, 11
Small
  agribusiness, 35
  and medium
    scale industries, 98
    sized businesses (SMEs), 21–24, 26, 29, 30
  industries, 37, 39, 47, 52
  scale
    companies, 139
    IT firms, 183
Smart
  industry, 91
  logistics, 29
Social
  activities, 221, 272, 282
  efficiency, 221
  enterprise, 218, 220, 222–231
    functionality, 224, 226
  entrepreneurial associations, 227
  entrepreneurs, 218, 221, 222, 225, 227
  entrepreneurship (SE), 217, 218, 220–224, 226–231
    organization, 222
  intelligence, 245
  intrapreneurship, 224, 229
  media, 22, 48, 57–65, 67, 68, 208, 210, 272, 276, 278, 280–283, 289–296, 298, 299, 313
    marketing (SMM), 289–291, 293–299
    platforms, 58, 60–63, 276
  mobility, 265
  networking, 290, 298
  networks, 62, 228, 290
  organizations, 225
  policy, 229
  sciences, 96, 205
  sustainability consequences, 25
  transformation, 225, 229, 230
  welfare, 265
Societal development, 218, 244
Socio-cultural
  conflict, 61
  system, 258
Socio-economic
  changes, 258
  development, 226, 229
  infrastructures, 177, 179
  life, 175
  scale, 243
  status, 243, 246
  value, 230
Socio-political
  activism, 223
  stability, 256
Soft cultural policy, 47
Software
  as a service (SaaS), 110, 114, 116, 121
  licenses, 115
  package for statistical analysis (SPSS), 40, 45, 79, 166, 243

programs, 63
requirements, 90
Solar energy, 183
Solid-state storage, 119
South America, 65, 69
Special accelerated road development program for north east (SARDP-NE), 177
Spinning vs. solid-state (SSD), 119, 120
Stakeholder, 67, 68, 78, 117, 140, 142–144, 213, 231, 264
communication, 22
expectations, 140
Standard operating procedure (SOP), 267
Standardization, 1
Start-up village entrepreneurship program (SVEP), 47
State plan assistance, 178
Statistical tools, 78
Strategic management, 186
Strategies, 22, 26, 40, 46, 48–53, 57, 58, 67, 69, 129, 141, 161, 241, 274, 290, 299
Structural inequality, 258
Structuralist approach, 258
Structured data, 313
Sub-unit decisions, 261
Supercomputers, 239
Supply
chain, 24, 25, 77, 211, 238
adaptability (SCA), 25
flexibility, 24
management, 23, 24
modularity, 25
chainwide real-time communication, 25
Sustainability, 22–27, 33, 85, 91, 127, 129, 140–143, 145, 146, 156, 162, 188, 217, 226–231, 263, 266, 268
values, 24
Sustainable
business practices, 141
design process, 188
development, 24, 25, 143, 176, 222, 226, 228, 229, 257, 269
functions, 25
goals (SDGs), 77, 143, 222, 223, 229
values, 24, 25
growth, 141, 175, 183
innovation, 24, 25
management, 74, 85
thinking, 24
tool, 291, 295, 299

T

Tax credits, 91
Tea tourism, 179
Technical-economic factors, 264
Technological
advancements, 23, 90, 95, 102
innovation, 23
Technology
adoption drivers, 30
benefits, 30
workshop center, 49
Telecom connectivity, 178
Telecommunication, 39, 130, 146, 182
Telemetry services, 130
Telepoint, 130
TensorFlow, 11
Terracotta, 36, 38
Text mining, 63
Textile, 38, 40, 51, 205
Thematic network analysis, 78
Third
industrial revolution (Industry 3.0), 88, 90, 91, 239
party
mediator, 108
service providers, 113
Three-dimensional (3D)
model, 230
printing, 89, 240, 304, 305
Threshold, 110, 134
Time-varying diversification ability, 6
Top-down approach, 260
Tourism sector, 177, 179
Traditional
crafts, 37, 47
currencies, 6
economic activities, 264
financial reporting, 143
freelancers, 46
women sector businesses, 35
Traffic
forecasting, 110
load balancing, 110
monitoring, 110

Transaction
costs, 2
processing systems, 62
volume, 127
Transactional
fees, 4
speed, 6
system, 5
Transparency, 118, 142, 152–154
Transport connectivity, 49
Transportation
links, 182
network, 179
Tribal
people, 37
population, 176, 182
Trickle-down effect, 257
Twitter, 59, 62, 64, 65, 69, 313
analysis, 59, 69

U

Udoji's policy making, 259
Uncertainty, 88, 100, 101, 275
Unconventional investment plans, 165
Uni-dimensionalization, 256
Unit root analysis, 131
United
Nations Capital Development Fund (UNCDF), 2, 74
States (US), 3, 4, 6, 207, 209
Dollar, 3, 4, 6
University
Grants Commission, 272
of South Africa (UNISA), 301
Unstructured data, 28
Urban
mobile subscriptions, 126, 128, 130–133, 135
transition, 164
Usage consistency, 128
User-friendly interfaces, 127

V

Vaccination, 74, 83, 84
Vaccine, 60, 64, 69, 76, 78, 83
Value-added services, 212
Variance inflation factor (VIF), 133
Varimax rotation, 148
Velocity, 62, 304, 306
Vendor, 75–77, 110, 116, 118, 193
development, 187
Veracity, 62, 304, 306
Vertical
system integration, 239
value chains, 89
Vigilant legal system, 61
Virtual
machines (VMs), 109, 119
RAM (vRAM), 119
reality, 28, 89, 187
service instance placement, 110
training, 28
Virtualization, 117
Vital healthcare services, 74
Vocal for local, 34, 37, 53
Vocational education, 38
Volatile market movements, 2
Volatility, 7, 8, 14, 88, 100, 101

W

Waste
management, 222
reduction, 26, 90
Water resources, 259, 264–266
management policies, 259
Wearable
computers, 203
devices, 201, 202, 206–209, 211, 212
electronics, 203
gadgets, 203
sensors, 208
technologies, 201–205, 207–211, 213
developments, 209
devices, 201
Web of science, 97
Wi-Fi networks, 202
Women
artisans, 36, 37, 40, 47, 50, 52
related strategy, 53
entrepreneurs, 33–35, 40, 48, 52, 53
Woodcraft, 36, 38, 46
Wooden
craft, 38
handicrafts, 33

Word
cloud, 66
analysis, 65
of-mouth marketing (WOM), 289
Work motivation questionnaire (WMQ), 243
World
economic development, 7
Health Organization (WHO), 59–61, 68, 77, 272, 273
Trade Organization (WTO), 2, 77
War II, 182